Advances in Information Security

Volume 81

Series editor
Sushil Jajodia, George Mason University, Fairfax, VA, USA

The purpose of the *Advances in Information Security* book series is to establish the state of the art and set the course for future research in information security. The scope of this series includes not only all aspects of computer, network security, and cryptography, but related areas, such as fault tolerance and software assurance. The series serves as a central source of reference for information security research and developments. The series aims to publish thorough and cohesive overviews on specific topics in Information Security, as well as works that are larger in scope than survey articles and that will contain more detailed background information. The series also provides a single point of coverage of advanced and timely topics and a forum for topics that may not have reached a level of maturity to warrant a comprehensive textbook.

More information about this series at http://www.springer.com/series/5576

Quanyan Zhu • Zhiheng Xu

Cross-Layer Design for Secure and Resilient Cyber-Physical Systems

A Decision and Game Theoretic Approach

 Springer

Quanyan Zhu
Electrical and Computer Engineering
New York University
New York, NY, USA

Zhiheng Xu
Electrical and Computer Engineering
New York University
New York, NY, USA

ISSN 1568-2633 ISSN 2512-2193 (electronic)
Advances in Information Security
ISBN 978-3-030-60253-6 ISBN 978-3-030-60251-2 (eBook)
https://doi.org/10.1007/978-3-030-60251-2

This Springer imprint is published by the registered company Springer Nature Switzerland AG
The registered company address is: Gewerbestrasse 11, 6330 Cham, Switzerland

To our families.

Preface

This book presents control and game-theoretic tools to design secure and resilient cyber-physical systems, including cloud-enabled control systems, 3D printers, robotic systems, and communication-based train control (CBTC) systems. We take a cross-layer perspective toward the design of these systems, as they are naturally composed of interdependent cyber and physical layers. The first part of the book (Chaps. 1 and 2) gives a general overview of the philosophy of the cross-layer design and its implications in smart cities, the Internet of Things, and industrial control systems. The second part of the book, which comprises Chaps. 3 to 5, focuses on securing cloud-enabled systems using customized cryptographic approaches. The applications of cloud-enabled systems are now ubiquitous in robotics, manufacturing systems, and distributed machine learning. This part includes extensive coverage of various techniques, including event-trigger control, homomorphic encryption, and customized encryption for outsourcing computation, in the context of dynamical systems.

Part III of the book presents game-theoretic approaches for cyber-physical system security and resilience. This part provides a range of applications, including 3D printers (Chap. 7), CBTC systems (Chap. 8), digital twins (Chap. 9), and robotic operating systems (Chap. 11), to demonstrate the application of game-theoretic tools. To help readers understand the analytical tools developed in these applications, Chaps. 6 and 10 give an overview of game theory and the partially observed Markov decision processes, respectively. Part IV of the book (Chap. 12) discusses the future research and challenges in this direction.

This book bridges multiple areas of research, including cryptography, control theory, game theory, and wireless communications. It contributes to the system-of-systems theory that provides a holistic approach for cyber-physical systems. Part II of the book has endeavored to establish the connections between cryptography and control theory. Cryptography is integrated into the control design, while the control theory informs the design of appropriate cryptographic solutions. The linkages between the two fields are central to providing fundamental security guarantees for cyber-physical systems. Part III of the book offers a unifying security and resiliency framework from the lens of game theory. Game theory is a mathematical science that investigates the strategic interactions among different agents or players. It can be used to describe the interactions between a defender and an attacker, a controller and disturbance, or interdependent networks. The observation that game-theoretic models can capture both cyber and physical interactions naturally leads to a holistic methodology for cross-layer analysis and design.

This book is useful as a supplementary textbook for a graduate-level course on recent advances in cyber-physical system security. Chapters 6 and 10, together with the materials in the appendices, provide necessary theoretical backgrounds for graduate students and practitioners. Other chapters in the book will provide readers a number of case studies that can motivate readers to solve similar problems at hand. This book is not meant to be comprehensive but aims to motivate readers to adopt a cross-layer system perspective toward security and resilience issues of large and complex systems. Readers are encouraged to collaborate with different domain experts and develop their methodologies that interconnect multiple research domains. In this way, we will witness a growing literature in this area and new exciting cross-disciplinary approaches to systems of systems.

Both authors would like to acknowledge their association with the Center for Cyber Security (CCS) and the Center for Urban Science and Progress (CUSP) at the New York University (NYU) when a major part of this book was completed. The offering of the special topics course on System Foundations for Cyber-Physical System Security and Resilience in Spring 2018 has motivated the need for textbooks and accessible reading materials in the area. The first draft of this book took shape after the second author completed his dissertation in 2017. Thanks to his persistence, we were able to finalize the draft in 2020. We both like to take this opportunity to thank many of our colleagues and students for their inputs and suggestions. Special thanks go to members of the Laboratory of Agile and Resilient Complex Systems (LARX) at the NYU, including Jeffrey Pawlick, Juntao Chen, Rui Zhang, Tao Zhang, Junaid Farooq, Linan Huang, Yunhan Huang, and Guanze Peng. Their encouragement and support have made this book possible. The first author would also like to acknowledge the funding support from the National Science Foundation (NSF) and the Army Research Office (ARO) for their support of this research.

New York, NY, USA   Quanyan Zhu
Singapore, Singapore   Zhiheng Xu
July 2020

Contents

Acronyms

AWGN	Additive white Gaussian noise
APT	Advanced persistent threat
AM	Additive manufacturing
AUV	Autonomous underwater vehicles
BIBS	Bounded-input bounded-state
CAD	Computer-aided design
CBTC	Communication-based train control
CCT	Cloud computing technology
CDP	Consecutive dropping packets
CE	Cloud-enabled
CPS	Cyber-physical system
DoS	Denial of service
DT	Digital twin
HMM	Hidden Markov model
ICS	Industrial control systems
ICT	Information and communication technology
IoT	Internet of Things
LMAC	Lightweight message authentication code
LSN	Large-scale sensor network
LQR	Linear quadratic Gaussian
LQR	Linear quadratic regulator
MAC	Message authentication code
MJS	Markov jump systems
MP	Markov partition
MPC	Model predictive control
MSE	Mean square error
NCS	Networked control system
NE	Nash equilibrium
PAR	Packet arrival rate
PBNE	Perfect Bayesian Nash equilibrium
POMDP	Partially observable Markov decision process

ROS	Robot operating system
RSA	Rivest-Shamir-Adleman (cryptosystem)
SCADA	Supervisory control and data acquisition
SG	Smart grid
SGE	Signaling game with evidence
SM	Subtractive manufacturing
SINR	Signal-to-interference-plus-noise ratio
SNR	Signal-to-noise ratio
STL	Stereolithography
UAV	Unmanned aerial vehicle

Symbols

$K^{m \times n}$	Set of $(m \times n)$-matrices with entries from set K
$A \times B$	Cartesian product of A and B
$\Pi_i A_i$	Cartesian product of sets A_i
2^V	Power set of V
$\Delta(X)$	Unit simplex over set X
$|X|$	Cardinality of set X
$\|\cdot\|_p$	General p-norm in $\mathbb{R}^n$, defined as $\|x\|_p = (\sum_{i=1}^n |x_i|^p)^{1/p}$, $p > 0$
$val(A)$	Saddle-point value of a game represented by payoff matrix A
$G = (V, E)$	Graph with node set V and edge set $E \subseteq V \times V$
$\mathcal{E}_C(\cdot)$	Encryption of a message under key C
$\mathcal{D}_C(\cdot)$	Decryption of a message under key C
x	Vector $x = (x_1, \cdots, x_n)$ as an element of set $X \subseteq \mathbb{R}^n$
x^T	Transpose of vector x
x_k	State vector x at time k
$\hat{x}_{k|k-1}$	Estimated state vector x at time k given the information at time $k - 1$
u_k	Control vector $u \in U$ from the admissible at time k
$\hat{x}_{k|k-1}$	Estimated control vector u at time k given the information at time $k - 1$
$\pi : I \to U$	Control policy that maps information sets I to the admissible set of control actions U
$\mathbb{E}(X|Y)$	Expectation of random variable X on Y
$cov(X|Y)$	Covariance between two random variables X and Y
$tr(X)$	Trace of matrix X
$\Pr(A|B)$	Probability of an event A given an event B
$\mathbb{E}_\pi(X|s_0)$	Conditional expectation of random variable X given the control policy π and initial condition s_0
$\text{argmin}_{x \in X} f(x)$	Sets of points in $x \in X$ where a minimum of objective function f is attained.

Part I
Motivation and Framework

Chapter 1
Introduction

1.1 Cyber-Physical Systems and Smart Cities

Recent years have witnessed an increasing interest in developing system technologies that integrate various Information and Communication Technologies (ICTs) with physical systems, ranging from large-scale infrastructures (e.g., transportation systems and power grids) to small systems (e.g., household appliances) [134, 184]. The integration of physical systems with cyber networks forms a Cyber-Physical System (CPS). It enables the use of digital information and control technologies to improve the monitoring, operation, and planning of the systems. This trend is pervasive in smart urban applications, including smart grids, connected and autonomous vehicles, and cyber manufacturing. For example, in a power system connected with networks, the operator can obtain the pollution concentration from remote sensors to regulate the power plant so that the concentration can stay at a healthy level. Another example is the Internet of Things (IoT). The cyber-physical nature of IoTs enables distributed sensing information in uncertain environments and real-time analytics for automating dynamic and adaptive decisions and control. In many applications, the integration with the cyberspace, e.g., the cloud, enables the outsourcing of massive computations to a third party, thus reducing the stringent hardware requirement to operate and hence making computations less costly and more efficient [177].

1.2 New Challenges in CPS

As CPSs have brought many promising benefits and advantages, they introduce new challenges that traditional methods are not sufficient to address. To understand the potential difficulties in CPSs, we need to explore the cyber-physical properties of

© The Editor(s) (if applicable) and The Author(s), under exclusive license to
Springer Nature Switzerland AG 2020
Q. Zhu, Z. Xu, *Cross-Layer Design for Secure and Resilient Cyber-Physical Systems*,
Advances in Information Security 81, https://doi.org/10.1007/978-3-030-60251-2_1

CPSs further. First, the performance of the physical layer of a CPS significantly relies on the capabilities offered by its cyber layer. For example, a multi-agent unmanned aerial vehicle (UAV) system requires that each agent exchange information with other agents through its cyber connections achieve a specific control objective, such as a formation or consensus-based control [175]. A reliable cyber communication plays a vital role in the control performance of Unmanned Aerial Vehicles (UAV). Disruptions in the communications caused by either nature hazards or human-made events will lead to failure of the planned mission and sometimes result in the UAV captured by an adversary [76].

Another example is Communication-Based Train Control (CBTC) systems, a class of automatic train control systems, integrated with wireless communications [147]. Each train can obtain its front train information through wireless communications. Given the front train's information, including velocity and position, the train can keep a safe distance from its front train. There have been many incidents of collisions due to either signaling failures or misinformation [45]. The control performance of the CBTC system relies heavily on the quality of communications enabled by the cyber layer.

Second, reversely, the physical layer's performance demands a performance guarantee at the cyber layer of the CPS. The design of the cyber layer needs to take into account the specifications required by the physical layer. Otherwise, it will be difficult, if not impossible, for the physical layer to achieve desirable performance. An example is the Networked Control System (NCS) where communication networks enable the channel between the sensors and the controller, and the one between the controller and the actuators. The design of stabilizing controllers is possible when the packet drop rates on these channels are higher than a fundamental threshold [70, 164, 205]. Otherwise, no stabilizing controllers are feasible. Hence the designs at the cyber-layer directly affect the feasibility of the designs at the physical layer. Hence, there is a need to impose constraints and specifications on the cyber-space designs to achieve a desired physical layer performance.

In summary, the main challenge of CPSs is that its strong interdependency between the cyber and physical layers provides opportunities for adversaries to inflict damage on the physical part of a CPS by infiltrating and compromising the cyber layer. This type of attack is distinct from traditional cyber attacks that aim only to take down cyber services. Instead, CPS enables cyber-physical attacks that can compromise physical assets through cyberspace. Take the CBTC system as an example again. A potential attacker can launch a jamming attack to disrupt the signaling between two trains. In this case, insufficient communications can cause severe consequences, such as emergency brakes and collisions between two trains. Stuxnet is a recent cyber-physical attack, which spreads as a computer virus that targets on Supervisory Control and Data Acquisition (SCADA) systems of an Industrial Control System (ICS) [109]. The Stuxnet intrudes the SCADA through its networked communication and disables the Programmable Logic Controllers (PLCs), leading to the malfunctioning of SCADA systems.

It has been observed that the traditional information security solutions, e.g., cryptography and intrusion detection systems, are insufficient to protect the CPSs from such attacks. On the one hand, cyber-physical attacks, such as Stuxnet, Operation Aurora, and Spyware, are examples of Advanced Persistent Threats (APTs) that leverage sophisticated attack techniques and can stealthily stay in the system for a long period without being detected. APTs are sophisticated and resourceful attacks that execute a sequence of cyber kill chain with specific objectives. Protection against these attacks cannot entirely rely on traditional techniques that often assume that adversaries do not know the cryptographic key or intrusion detection algorithms. On the other hand, traditional methods mainly protect cyber assets. Once the attacker successfully compromises the cyber assets and reaches the physical layer, there are no other defenses that can deter the attacker from disrupting the physical systems or allow the CPS to maintain its function. Hence, there is a need to develop new mechanisms to mitigate the impact of cyber-physical attacks on CPSs.

This book focuses on two aspects of the mechanisms. One is the security mechanism that aims to defend the cyber layer from APT-type attacks and deter the attacker from reaching the physical layer. The other is the resiliency mechanism that allows the physical layer to maintain its performance and recover from the failures even when a successful attack is launched. The security mechanism focuses on hardening the cyber layer's security, while resiliency mechanisms focus on mitigating impacts on the physical layer's performance.

The interdependency between the cyber and the physical layers of CPS naturally makes these two aspects of the mechanism intertwined. Resiliency relies on security, while security serves resiliency. There is a need for a holistic view to address both CPS design challenges. This book introduces a cross-layer approach that integrates both design objectives and develops mechanisms for secure and resilient CPS. To illustrate our methodology, we use case studies in multiple areas, including UAV systems, large-scale sensing networks, networked 3D printers, CBTC systems, and robotic systems. Each case study aims to address a specific connection between the cyber and physical layers of the CPS and present a particular set of modeling and design tools. We divide the cross-layer design methodologies into two parts. The first one bridges the gap between cryptography and control-theoretic tools. It develops a bespoke crypto-control framework to address security and resiliency in control and estimation problems where the outsourcing of computations is possible. The second one bridges the gap between game theory and control theory and develops interdependent impact-aware security defense strategies and cyber-aware resilient control strategies.

1.3 Overview and Related Works

The issues investigated in this book are related to the security issues in cyber networks. The traditional cryptography and information technologies focus on data confidentiality and integrity. One application is cloud outsourcing computation. Recent years have witnessed significant growth in this field. Recent works have developed secure protocols for outsourcing scientific computations, such as linear programming problems [170], large matrix inversion problems [87], large matrix determinant problems [88]. Cong et al. have developed cryptographic mechanisms to achieve confidentiality and integrity in the outsourcing computation. However, CE-NCSs have unique cyber-physical features because their physical control components are tightly integrated with other counterparts. The protocols designed for outsourcing computation are not sufficient to address these features.

Another type of CPSs is the cloud robotics and automation [75], which investigates object recognition, robot navigation, and path planning. These are practical examples of CE-CPSs. Other works have also developed applications using the architecture of CE-CPSs to enhance efficiency, e.g., cloud-based transportation system [68] and cloud-based grasping system [74]. Nevertheless, very few works have provided a fundamental understanding of CPSs. Besides, security challenges are often ignored in the design of these systems.

The conventional IT technologies are not sufficient to address these challenges in CPSs since the cyber layer of a CPS is coupled with its physical dynamics [146]. From a control system's perspective, we can view CPSs as networked control systems (NCSs), and many attack models on NCSs have been discussed in the literature [165], including Denial-of-Service (DoS) attacks [124], replay attacks [108], and data injection attacks [96]. We will also analyze some of them in specific applications.

Many methods have been designed to deal with cyber-physical security issues. One critical approach is to apply game-theoretic tools. Game theory has shown promise in a wide range of security issues as it can capture and model the strategic interactions between the attackers and defenders [65, 101]. In [201, 204], Zhu et al. have applied a game-theoretical framework to improve the resilience of CPSs in the face of attacks. The framework constructs two game models to capture the cyber-physical nature of CPSs. The physical-layer game framework models the interactions between the physical system and disturbance, while the cyber-layer game framework captures the interactions between an attacker and a defender.

In Chap. 10, we study the security issues of networked Robot Operating Systems (ROSs) [132]. Hence, the work in Chap. 10 addresses emerging security challenges for ROSs. In [103], McClean et al. have analyzed common, low-cost, low-overhead cyber attacks on a ROS. Given these increasing threats to the ROSs, Bernhard et al. have implemented standard encryption and authentication to achieve data confidentiality and integrity in the ROS-based applications [41]. In this book, we will address the cyber-physical security challenges of ROSs and focus on designing optimal security mechanisms that consider the tradeoffs between the performance of delay-sensitive control systems and the message integrity of ROSs.

1.4 Outline of the Book

This book consists of four parts. Part I of the book discusses the motivation and the cross-layer framework for CPS. Part II presents the cryptographic and control-theoretic approach as the cross-layer design paradigm for cloud-enabled CPS. Part III presents a game-theoretic foundation for secure and resilient CPS design. Part IV concludes the book and presents future research directions. The remaining chapters of the dissertation are organized as follows.

Chapter 2 presents the underlying architecture of the cross-layer design to achieve secure and resilient control of CPSs. Then, we show three branches of cross-layer design: the first one uses cryptographic methods to attain secure outsourcing computations of a CPSs; the second one uses game-theoretical tools to realize safe and reliable control of critical CPSs; the third one uses partially observed Markov decision process to study an incomplete information case.

Chapter 3 presents the architecture of the cloud-enabled CSPs. This architecture brings many benefits to CPSs, but introduces new challenges when the cloud is untrusted. In Chap. 4, we develop a secure and resilient outsourcing protocol for cloud-enabled CPSs. In Chap. 5, we combine the standard and customized homomorphic encryption to establish an outsourcing data assimilation for large-scale sensor networks.

Chapter 6 reviews three game-theoretic models and discusses how these game models can contribute to the cross-layer design. In Chap. 7, we use *FilpIt* game to develop a security mechanism to protect 3D printers from APTs. In Chap. 8, we use a zero-sum stochastic game to achieve a secure control of Communication-Based Train Control (CBTC) systems.

Chapter 9 introduces Partially Observed Markov Decision Processes, including their structural properties and algorithmic solutions. In Chap. 10, we use POMDPs to develop a secure and resilient mechanism to protect the Robot Operating Systems (ROSs) from false-data-injection attacks.

Chapter 11 discusses the future directions in the secure and resilient design of CPSs. The increasingly intelligent and sophisticated attacks create numerous new challenges in CPSs.

Chapter 2
Cross-Layer Framework for CPSs

2.1 Introduction to Cross-Layer Design

The architecture of CPSs can be hierarchically divided into cyber and physical layers. The cyber layer consists of cyber assets, such as servers, workstations, sensors, wired or wireless communications, and networking. In the physical layer, the physical assets include actuators, controllers, and plants. Figure 2.1 illustrates the hierarchical structure of CPSs. The cyber layer represents cloud computing, which includes communication networks and servers for computations and storage. Infrastructures, such as power systems, autonomous vehicles, buildings, and public health, are examples of the physical assets that need service from the computing facilities at the cyber layer.

Many legacy systems in critical infrastructures are often isolated from the networks in the past, and it is difficult for an attacker to observe or access the sensitive information of the physical system. However, with the rapid development of CPSs, the integration with information and communication technologies exposes the physical data and equipment to cyber attacks. Traditional IT solutions are not sufficient to protect the CPSs from these cyber attacks. One reason is that IT solutions mostly focus on data confidentiality and integrity in the cyberspace. For CPSs, data availability is more essential since most of the control systems are time-critical; i.e., they require real-time data to stabilize the system. Another reason is that the standard assumption that the secret key cannot be obtained by the attackers may not hold anymore. Nowadays, attackers can launch sophisticated attacks, e.g., Advanced Persistent Threats (APTs) [39], that can go through multiple stages of reconnaissance, weaponization, exploitation, and command and control to achieve the attack objectives. They can often stay in the system for a long period of time and obtain secret keys or information before moving to the next step in the kill chain.

Another challenge of safeguarding CPSs arise from the interdependencies of the cyber and the physical layer systems. The goal of CPS security is not just to defend

© The Editor(s) (if applicable) and The Author(s), under exclusive license to
Springer Nature Switzerland AG 2020
Q. Zhu, Z. Xu, *Cross-Layer Design for Secure and Resilient Cyber-Physical Systems*,
Advances in Information Security 81, https://doi.org/10.1007/978-3-030-60251-2_2

Fig. 2.1 Examples of
cyber-physical systems

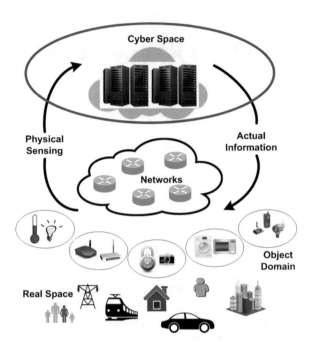

the cyber system but also to protect the physical system. The design at the cyber layer affects the performance of the physical layer. The design at the physical layer demands a service-level guarantee from the cyber layer. The two layers have to be designed together in a holistic manner to improve the security at the cyber layer and the resiliency at the physical layer. To this end, this book aims to introduce a cross-layer approach. The cyber-physical interactions in an adversarial environment can be conceptually represented by Fig. 2.2, where x denotes the physical state of the system; θ denotes the cyber state; u represents the physical layer control; l represents the defense strategy at the cyber layer. In an adversarial environment, an adversary launches an attack a through the cyber system to compromise the physical plant, which is subject to natural disturbances or uncertainties w.

For cyber systems, threat models are useful to describe the way how an attacker interacts with a defender. For physical systems, domain-specific physical modeling describes how a physical plant is controlled to achieve desirable performances. In the cross-layer design framework, the controller l is designed to be aware of the cyber performance, while the defense strategy d is designed to be aware of its impact on the physical layer performance. Depending on the specific nature of the problem and the design criteria, the cyber and the physical models can take different forms and require different design methodologies. In the following, we introduce two specific cross-layer design frameworks. One relies on the connections between the cryptography with control theory, where encryption and decryption mechanisms are designed together with the controller. The other one leverages game theory to unify the modeling of threat vectors and control systems. The unified game-

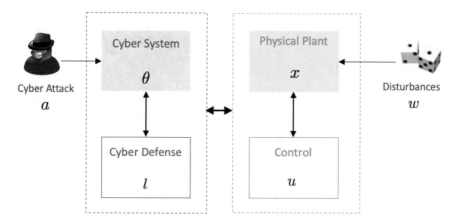

Fig. 2.2 Conceptual framework of the interactions between the cyber and the physical layers of the system under adversarial environment

theoretic approach leads to a natural games-in-games framework for cyber-physical co-designs.

2.2   Cross-Layer Design: Connecting Cryptography and Control Theory

The security issue at the cyber layer may take different forms in Fig. 2.2. In this section, the goal at the cyber layer is to provide data privacy and integrity leveraging cryptographic techniques; i.e., l in Fig. 2.2 represents a cryptographic design. The goal at the physical layer is to ensure system stability; i.e., u in Fig. 2.2 represents a stabilizing control design. There is an interplay between cryptographic and control designs. The cryptographic solution l may lead to computational overhead and signaling delays for the stabilizing control u. Hence standard techniques, such as public-key cryptography and homomorphic encryption, are not directly applicable to CPSs. There is a need for a customized cryptographic design that takes into the control performance requirements, and a bespoke control design that is aware of the delay and overhead created by the cryptographic design.

In Chaps. 4 and 5, we provide two specific applications of cloud-enabled CPSs (CE-CPSs). Figure 2.3 illustrates the architecture of a CE-CPS. A control system is connected to the cloud, where the heavy computation in the controller is outsourced to the cloud, and the computed result is fed back to the controller. In this way, the controller can offload massive computation to the cloud, thus improving the computational efficiency. The cloud may not be trusted. There is a need to develop a mechanism that can achieve data confidentiality, integrity, and availability.

Fig. 2.3 Architecture of a
CE-CPS: The integrated
system consists of a plant,
sensors, actuators, and a
cloud-based controller. The
cloud-based controller is
composed of a local
controller, a wireless network,
and a cloud. The network is
deployed to transmit data of
computations to the cloud and
the control solutions from the
cloud. The controller aims to
stabilize the plant and achieve
system objectives

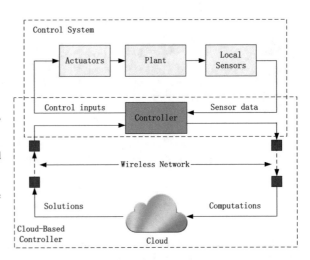

2.3   Cross-Layer Design: Connecting Game Theory with Control Theory

Designing cryptographic solutions is often not sufficient for sophisticated attacks,
such as APTs. Instead of relying on them, we need to find other security mechanisms
to defend the systems strategically, given limited resources. Examples of such
mechanisms include moving target defense [71, 203], cyber deception [72, 128],
and network randomization [5]. Game theory provides a suitable mathematical
framework for understanding strategic decision-making. It is a natural framework to
model adversarial environments in which an attacker interacts with a defender with
conflicting objectives [101]. Apart from the attacker's capabilities, game theory
provides a richer model that can capture the incentives, the long-term behaviors, and
the information structure of the adversary. In a game-theoretic model, a in Fig. 2.2
represents the attacker's strategy, while l corresponds to the defender's strategy.
Interestingly, the interactions between the controller and the disturbance can also be
interpreted as a game. The controller u aims to optimize the control performance
with minimum control effort, while disturbance w can be viewed as one that has
an opposing objective. Such controller design paradigm is called H_∞ or optimal
robust control design [11]. The game-theoretic approach unifies the cyber and the
physical layer designs since the design problems can be both viewed as games
of different nature. This observation is illustrated in Fig. 2.4. The physical layer
interactions between u and w are captured by the physical system game (PSG),
while the cyber layer interactions between a and l are captured by the cyber system
game (CSG). Hence the cross-layer design can be viewed as a game-in-games, or a
meta-game [32, 65, 204], where two games are clearly interdependent. The meta-
game modeling can thus drive the cross-layer design.

Fig. 2.4 Game-theoretic approach to CPS modeling. The interactions between the disturbance and the controller at the physical layer are captured by the physical system game (PSG). The interactions between the defender and the attacker at the cyber layer are captured by the cyber system game (CSG). The two games interact and form a meta-game

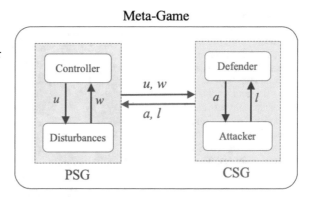

In Part II of the book, we use multiple case studies to illustrate this design philosophy. Figure 2.5 illustrates the general architecture of a CPS. At the cyber layer, we use a game model to capture the interactions between attacks and defenders. At the physical layer, we use another game model to describe the behaviors between the controller and the disturbance. The cyber game aims to find the best solution to mitigate the system's impact, while the physical game provides a robust controller to stabilize the physical systems. Normally, the physical layer requires the reference or feedback information from the cyber layer. Hence, physical actions are based on the actions of the cyber layer. One way to compose the two games together is to use a leader-follower structure, i.e., the Stackelberg game framework. The physical layer is the follower who makes a decision based on the observation of the cyber actions; the cyber layer is the leader who makes a decision based on the anticipation of the physical actions. Thus, we build a connection between these two layers and find an optimal solution to protect the system. In Chap. 7, we use 3D printers as an application to illustrate our design.

2.4 Cross-Layer Design Under Incomplete Information

In the cross-layer design framework, we can use different models to capture the prosperities of different layers. It has been assumed in Fig. 2.2 that the cyber state θ or the physical state x are completely observable. However, in many applications, cyber states are not directly observable. For example, illustrated in Fig. 2.6, we define two cyber states. The working state means that the cyber layer works as expected, while the failure state states otherwise. Based on the actions of the defenders or attackers, we use a Markov decision model to describe the dynamics of the cyber state. In Fig. 2.6, parameters γ and ρ are the transition probability, and a is the action taken by the defender.

Obtaining the full information of the cyber state is difficult. Hence, the MDP cannot capture the property of incomplete knowledge. To address the issues, we

Fig. 2.5 The
Games-in-Games
Architecture: we use a
`FlipIt` game to capture the
interactions between defender
and attacker; we develop a
zero-sum game to describe
the relationship between the
controller and disturbance;
we use Stackelberg game to
capture sequential behaviors
between cyber and physical
layers

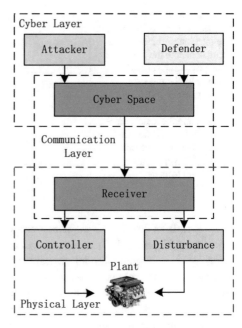

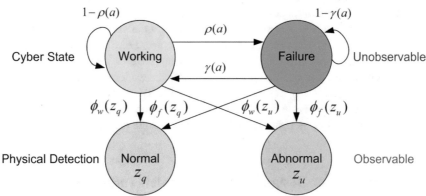

Fig. 2.6 The architecture of the incomplete information model: the defender cannot directly
observe the cyber state, but it can observe the outcome of the physical detector

develop a physical detector to diagnose the abnormal situation. The distribution
of detection results depends on the cyber state. Given the detector, we construct
a belief of the cyber state. In Fig. 2.6, ϕ_w and ϕ_f are the detection rates, which
are functions of the detection results. z_q and z_u are the outcomes of the physical
detector. Chapter 10 introduces partially observable Markov decision processes
(POMDPs) to provide a system design framework under incomplete observations.
We will show in Chap. 11 that POMDPs can be applied to design security and
resiliency mechanisms for robotic operating systems (ROS).

2.5 Conclusions

In this chapter, we have introduced a conceptual framework for the cross-layer CPS designs in which the cyber system and the physical systems are designed jointly to achieve desirable interdependent performances. We have presented two specific frameworks for secure and resilient control of CPSs. The first one leverages cryptographic solutions to enable data privacy and integrity, and control-theoretic tools to develop a controller to stabilize the physical systems. The cross-layer framework bridges the two design paradigms by understanding the confluences between cryptography and control theory. The second one presents game theory as a unifying framework for designing cyber defense strategy in adversarial environments and achieving optimal control design under worst-case disturbances. The unified perspective enables a games-in-games approach to develop security and resiliency mechanisms for CPSs. We have shown that these two frameworks can be extended to those under which the states are not perfectly observable. In the ensuing chapters, we will illustrate the cross-layer design methodologies using specific applications, including autonomous systems, 3D printers, and Robot Operating Systems (ROS).

Part II
Secure Outsourcing Computations of CPS

Chapter 3
New Architecture: Cloud-Enabled CPS

3.1 Promising Applications of CE-CPSs

The advent of Cloud Computing Technologies (CCTs) has revolutionized the development of CPSs. The integration of CCTs with CPSs leads to Cloud-Enabled CPSs (CE-CPSs), bringing the following revolutionary features to control systems: large-scale real-time data processing, massive paralleled computation, and signal processing, and resource sharing. Another advantage of outsourcing complicated computations to a cloud is that the local system can reduce its energy cost, enabling the system to conduct a mission in a longer duration.

Given the benefits of CCTs, we will see a growing number of CE-CPSs' applications, varying from different areas, e.g., robotics, smart grids, intelligent transportation systems, and modernized manufacturing. The main common property of these applications is that they require massive computational abilities, big data-sharing, and critical real-time performance. The cloud-enabled architecture can achieve these objectives efficiently and effectively. In the following subsections, we will briefly introduce these applications and outline the essential references for the readers.

3.1.1 Cloud-Enabled Robotics

Robots have improved our lives in many ways over the past decades. They contribute varied solutions to real-world problems, such as extra-operations, automated manufacturing, unmanned rescue and search, self-driving vehicles medical robots [75]. Among these applications, the growing complexity of requirements and tasks introduces computational issues for robots, since many of them are using embedded

© The Editor(s) (if applicable) and The Author(s), under exclusive license to   19
Springer Nature Switzerland AG 2020
Q. Zhu, Z. Xu, *Cross-Layer Design for Secure and Resilient Cyber-Physical Systems*,
Advances in Information Security 81, https://doi.org/10.1007/978-3-030-60251-2_3

Fig. 3.1 The Cloud Robotic and Automation: the resource-limited robots can outsource their massive computations to a cloud, improving their computational efficiency

systems with limited computing resources [62]. Besides, the locally available information also constrains individual networked robots' performance, which can only access the resources accumulated by the local sensors.

To address the above issues in robotics, researchers have proposed Cloud-Enabled Robotics. The new cloud-enabled framework utilizes the elastic on-demand resources offered by a ubiquitous cloud infrastructure [144]. Figure 3.1 illustrates a variety of applications in cloud robotics. Arumugam et al. [8] have proposed a software framework, named DAvinCi that provides scalable advantages for service robots in large environments. Another related application is the RoboEarth project [138], driven by an open-source cloud robotics platform, named Rapyuta [69]. The platform enables the robots to outsource their heavy computations to the cloud through networked communications. Rapyuta provides a compactly interconnected computing environment, making it efficient for the deployment of multi-agent robotic systems.

3.1.2 Cloud-Enabled Smart Grids

mart Grids (SGs) play a significant role in electric power utilities, which is essential to our daily lives. SGs can facilitate the monitoring and control of power usage in a large scale area. Hence, the fundamental requirements for advanced SGs include reliable and secure communications, efficiency computations, small delay, and information sharing. For instance, the SG should balance and manage its energy

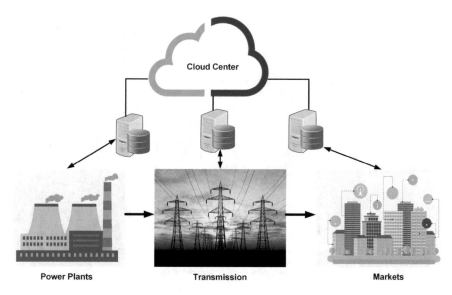

Fig. 3.2 The Architecture of a CE-SG: the CE-SG performs like a huge feedback system, in which the cloud center assembles all the critical information of the entire SG and provides optimal solutions to each individual system

load when the consumed energy reaches its peak levels [154]. The energy-balanced problem remains non-trivial in a large-scale SG.

The new computing paradigm, Cloud-Enabled SGs (CE-SGs), can address the issues mentioned above. CCTs offer massive computational services with its big data centers for storage. Besides the powerful computational ability, CCTs can also facilitate communication and data-sharing among large-scale SGs. Many studies have already focused on CE-SGs. Simmhan et al. [153] have studied how CCTs can optimize the demand response of a large-scale SG. Given the CE-SGs, they introduce a flexible and scalable cloud virtual machine model, which allows the SGs to have more freedom in choosing available resources to respond to the incoming demands. The work in [78] introduces another unique CE-SG architecture, which facilitates real-time data retrieval and parallel processing for SG applications. Readers can find many other CE-SG related applications in literature [15, 20, 59]. Figure 3.2 shows a structure of CE-SGs, which can assemble all the information and optimize the solutions to the entire smart grids.

3.1.3 Cloud-Enabled Transport Systems

Intelligent Transport Systems (ITSs) are a set of advanced applications that aim to use intelligent and big-data technologies to conduct innovative service for traffic management and transport [42]. Cloud-Enabled Intelligent Transport Systems (CE-

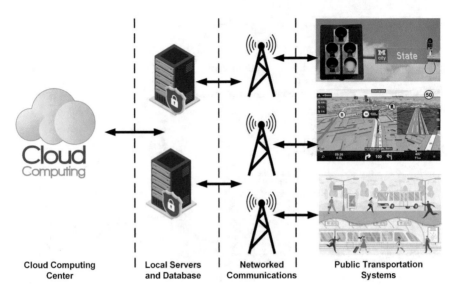

| Cloud Computing | Local Servers | Networked | Public Transportation |
| Center | and Database | Communications | Systems |

Fig. 3.3 The Architecture of a CE-ITS: Based on the real-time traffic data, the cloud can optimize the solutions to the traffic issues and provide intelligent decisions to the public transportation systems

ITSs) can meet the above requirements by providing large amounts of storage and computing resources. Besides, CCTs can offer services, such as decision support and a standardized development environment for transportation management.

The advantages of CE-ITS have attracted many researchers to develop a new framework for modern transport systems. Bitam et al. [19] have proposed an ITS-Cloud framework to improve transportation performance, such as road safety, transport productivity, travel reliability, environment protection, and traffic resilience. Li et al. [91] have analyzed how cloud computing can contribute to agent-based urban transportation systems. Figure 3.3 illustrates a basic structure of the CE-ITS, which can use the cloud to optimize the performance of the transportation system.

3.1.4 Cloud-Enabled Manufacturing

Given the concept of Industrial 4.0, researchers and operators have aimed to drive the manufacturing toward challenging directions, such as the fastest time-to-market, highest equality, lowest cost, best service, and greatest flexibility [161]. The new architecture, Cloud-Enabled Manufacturing (CE-Mfg), can efficiently achieve those objectives. A CE-Mfg is a computing and service-oriented manufacturing model developed from the existing advanced manufacturing model. CE-Mfg's primary goal is to realize the full sharing and circulation, high utilization, and on-demand use of numerous manufacturing resources and capabilities [162]. Given these advantages, CE-Mfg has already been deployed in a group of critical manufacturing areas, such as IT, pay-as-you-go business models, production scaling up and down per demand, and flexibility in deploying and customizing solutions [173].

3.2 New Security Requirements of CE-CPSs

Despite many advantages, new challenges arise in CE-CPSs, since the cloud is untrusted. When the CPSs outsource their computations to a cloud, they need to send necessary data to the cloud. A potential attacker can intrude the cloud to steal, modify, or disrupt the data transmitted by the CPSs. Since CPSs are sensitive to the feedback data, these attacks might lead to severe damages to the system. Given the cyber threats, we need to design a secure outsourcing protocol to protect the CE-CPSs. We propose the following objectives of the security mechanism for CE-CPSs. For example, Fayyaz et al. have discussed numerous security issues in CE-SGs [49].

Given the new security challenges in CE-CPSs, we aim to design security protocols to protect the systems from cyber attacks in the following Chaps. 4 and 5. Here, we outline the generic objectives of the proposed mechanism. The main objectives are summarized as:

- **Confidentiality**: No sensitive information in the original control problems can be obtained by the cloud server or adversaries during the computation of the problem.
- **Correctness**: Any cloud server that faithfully runs an application should generate an output that can be decrypted to a correct solution for the control system.
- **Integrity**: The correct results from a faithful cloud should be verified successfully by using a proof from the cloud. No false result can succeed in the verification.
- **Efficiency**: The local computation (e.g., encryption, verification, and decryption) processed by the control system should be less heavy and cheaper than that of solving the original problem.
- **Resiliency**: The control system can maintain stability when no verifiable solutions are available from the cloud. The control system can recovery online after an *availability attack*.

3.3 Conclusions

Integrated with cloud computing, CE-CPSs have a promising future in many application domains. In this chapter, we have introduced CE-CPS's architecture and several potential applications. We have discussed that the cyber-physical nature of the system introduces new challenges to the system. In Chaps. 4 and 5, we will present two specific applications of CE-CPSs. We present a cross-layer approach to secure outsourced data processing and improve the resiliency of the control systems. We will demonstrate how these mechanisms can achieve the design objective presented in Sect. 3.2.

Chapter 4
Secure and Resilient Design of Could-Enabled CPS

4.1 New Challenges and Proposed Solutions of CE-CPS

Feedback control systems are ubiquitous in modern critical infrastructures (e.g., power grids and transportation) as well as in civilian lives (e.g., elevators and robots). In the classical design of control systems, all system components, including sensors, controller, actuator, and plant, are embedded in a local unit. This practice leads to a high cost of installation, communication constraints, and a lack of flexibility. The advent of Information and Communication technologies (ICT) greatly facilitates the integration of advanced technologies into control systems. Feedback loops are integrated with wireless communications between different system components as CPS, which enables a broad range of modern control system applications, such as remote surgery [105], Unmanned Aerial Vehicles (UAVs) [148], and smart grids [53].

Figure 4.1 illustrates an example of UAV that conducts a searching mission. As the UAV is resource-constrained, it can improve its performance by outsourcing the computations of the searching and control problems to the cloud. One real application of CE-CPSs is the Google self-driving car [75]. The self-driving car, connected to satellites to collect and update the maps and images, can upload the sensing information to a cloud to achieve accurate positioning. Another example is the cloud-based formation control [166]. The authors have demonstrated that the integrated robotic system with a cloud simplifies the hardware of the robots as the robots outsource heavy computations to a cloud. In [121], Pandey et al. have developed a dynamic mechanism to integrate resource-constrained networked robots with cloud, and Autonomous Underwater Vehicles (AUVs) are used as an example to demonstrate the benefits of collaboration between the AUVs and cloud resources.

Q. Zhu, Z. Xu, *Cross-Layer Design for Secure and Resilient Cyber-Physical Systems*, Advances in Information Security 81, https://doi.org/10.1007/978-3-030-60251-2_4

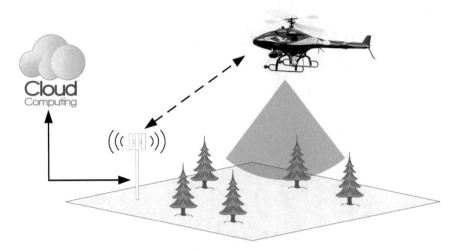

Fig. 4.1 An example of the Cloud-enabled CPS: an unmanned helicopter conducts a search mission and outsources its computations to a cloud

Despite the advantages of CE-CPSs, new security challenges arise from both the cyber and the physical parts of CE-CPSs. The first challenge is the confidentiality and integrity of the data transmitted between a cloud and CPSs. One the one hand, outsourcing computations to a cloud may expose the private information (e.g., parameters of the system model and sensor information) of the control system to cyber attackers [170]. On the other hand, cyber adversaries can modify the data between the cloud and the control system to disrupt data integrity. The second one comes from the control system, which requires feedback control inputs at every sampling time to guarantee stability. The lack of control inputs can lead to instability of CPSs. The third one is efficiency. The amount of local computation, including encryption, decryption, and verification, performed by the control system, should be less than that of solving the original control problems [87].

To address these challenges, we propose a secure and resilient control design for CE-CPSs. In our design, the CPSs encrypt the data prior to the computation in cloud, and verify the solutions from the cloud. The proposed encryption differentiates standard encryptions as the encryption and decryption are only processed on the control system. This removes the expensive public key exchange in standard encryption. To guarantee the stability, and enhance its resilience to cyber attacks, we design a Switch Mode Mechanism (SMM) with three working modes: Cloud Mode, Buffer Mode, and Safe Mode. The controller of the CE-CPSs can switch between these modes when unanticipated events occur.

4.2 Problem Statements

Each CE-CPS has two layers: the cyber layer and the physical layer. The cyber layer consists of wireless communications and a cloud, while the physical layer incorporates a plant, actuators, and sensors, and a controller. The integration of the controller with the cloud constitutes a cloud-based controller. Figure 2.3 illustrates a feedback architecture of a CE-CPS. The controller of a CE-CPS is structurally different from other general CPSs. Instead of solving an offline control problem, the controller of a CE-CPS formulates a dynamic optimization problem based on the sensor data and outsources the computations of the control decisions to a cloud. After receiving the solutions from the cloud, the controller sends them to the actuators of the physical system.

To describe in detail the architecture of a CE-CPS, we first introduce the physical layer control problem. In this work, we present a discrete-time linear system to capture the dynamics of the physical plant and use Model Predictive Control (MPC) framework to design optimal control to stabilize the system. The computations of the MPC control inputs are outsourced to the cloud, which can be subject to adversarial attacks. To this end, the second part of this section presents three attack models on the CE-CPS, and outline the design goals and framework of the proposed mechanism.

4.3 System Dynamics and MPC Algorithm

MPC has been widely used in many domains of industries and civil applications, such as process control of chemical plants and oil refineries, energy systems in buildings, and autonomous vehicles [7, 185]. MPC is a model-based control strategy that uses a system model to predict the future behaviors of the system to establish appropriate control inputs [168]. To achieve prediction, MPC strategies provide a moving finite-horizon problem based on the system model, and control inputs can be computed by solving this problem at every sampling instant. The periodical property of MPC makes it possible to handle constraints on control inputs or system states [77], and this advantage makes MPC more practical in many applications.

To apply MPC, we use a discrete-time linear system model to describe the system dynamics of a control system, given by

$$x_{k+1} = Ax_k + Bu_k, \qquad (4.1)$$

where $x_k \in \mathbb{R}^n$ is the state vector of the CPS, $x_0 \in \mathbb{R}^n$ is the given initial state value, $u_k \in \mathbb{R}^l$ is the control input, $A \in \mathbb{R}^{n \times n}$ and $B \in \mathbb{R}^{n \times l}$ are constant matrices.

At each sampling time k, a finite-horizon MPC problem $\mathcal{P}$ is a minimization problem, which is given as follows:

$$\mathcal{P} : \min_{U_k} J(x_k, U_k) = \sum_{\tau=0}^{N-1} \left(\|\hat{x}_{k+\tau|k}\|^2 + \eta\|\hat{u}_{k+\tau|k}\|^2 \right), \qquad (4.2)$$

subject to

$$\hat{x}_{k+\tau+1|k} = A\hat{x}_{k+\tau|k} + B\hat{u}_{k+\tau|k},$$
$$\hat{x}_{k|k} = x_k, \ \hat{u}_{k+\tau|k} \in \mathcal{U},$$
$$U_k = [\hat{u}_{k|k}^T, \hat{u}_{k+1|k}^T, \ldots, \hat{u}_{k+N-1|k}^T]^T,$$

where $J : \mathbb{R}^{n \times 1} \times \mathbb{R}^{lN \times 1} \to \mathbb{R}$ is the objective function, η is a tuning parameter for the desired control performance, and $U_k \in \mathbb{R}^{lN \times 1}$ is the solution sequence of the problem $\mathcal{P}$. $\hat{x}_{k+\tau|k}$ denotes an estimate value of $x_{k+\tau}$, based on the feedback state x_k.

Due to the existence of noise and disturbances, an error between $\hat{x}_{k+\tau|k}$ and x_k always exists. For this reason, the CPS solves the problem $\mathcal{P}$ for each k, and only takes the first element $\hat{u}_{k|k}$ of U_k as the control input at time k.

4.4   The Standard Form of Quadratic Problem

It is convenient to transform $\mathcal{P}$ into a standard quadratic form. The optimization problem (4.2) becomes

$$\min_{U_k} J(x_k, U_k) = \hat{X}^T \hat{X} + \eta U_k^T U_k = (\Gamma x_k + HU_k)^T (\Gamma x_k + HU_k) + \eta U_k^T U_k.$$

Since the constraints are often given in the algebraic forms of inequalities and equalities, we can transform the constraints $\hat{u}_{k+\tau|k} \in \mathcal{U}$ into the following inequality and equality constraints, $MU_k \le c$, $EU_k = d$, where $M \in \mathbb{R}^{m \times lN}$, $E \in \mathbb{R}^{p \times lN}$, $c \in \mathbb{R}^{m \times 1}$, and $d \in \mathbb{R}^{p \times 1}$ are constant matrices and vectors. Then, we define

$$Q = H^T H + \eta I, \ b^T = 2x_k^T \Gamma^T H, \qquad (4.3)$$

where I is an $lN \times lN$ identity matrix, the size of Q is $lN \times lN$, and the length of vector b is lN. By eliminating the terms independent of the U_k, we transform $\mathcal{P}$ into a standard quadratic form:

$$\mathcal{QP} : \min_{U_k} \ J(x_k, U_k) = U^T(k)QU_k + b^T U_k$$

$$\textbf{s.t.} \ \ MU_k \le c, \ EU_k = d.$$

The $\mathcal{QP}$ can thus be defined by the tuple $\Psi := \{Q, M, E, b, c, d\}$.

4.4.1 Cloud Attack Models

The security threats mainly come from the cloud and the networked communications between the cloud and control system. To capture the possible behaviors of adversaries, we present three attack models:

(a) **Ciphered-only attack**: The attacker observes the encrypted information between the cloud and the control system, and attempts to determine the original information. For example, the attacker can simply record all the information it can access in the cloud, and use this to learn the sensor values and control inputs that should remain private [87].

(b) **Message modification attack**: The attacker may either intrude the cloud or use spoofing techniques [76] to modify the solution that is sent to the control system. The fake solution can mislead or disrupt the control system, leading to catastrophic consequences.

(c) **Availability attack**: The attacker blocks the communication between the cloud and the control systems so that the control systems cannot receive correct solutions from the cloud for either a short or long duration. As control systems require the feedback control inputs to stabilize themselves, this type of attack can lead to instability of the systems.

4.4.2 The Framework of the Proposed Mechanism

Figure 4.2 illustrates the structure of the framework. At every sampling time k, the control system first formulates a control problem Ψ, encrypts it with a key K to obtain Ψ_K, and sends it to the cloud. The cloud runs **ProbSolve** to solve Ψ_K, and returns a solution y together with a proof Ω. The CPS verifies y with Ω. If the verification succeeds, the algorithm **ResVer** output $\Pi_v = 1$, and the CPS decrypts y to obtain the control input U_k. If the verification fails ($\Pi_v = 0$), then the CPS switches to a Buffer Mode, and checks a switching condition Π_s to determine whether the CPS needs to switch to a Safe Mode. These modes will be designed later in parts.

4.5 Confidentiality and Integrity

The first two attack models, *ciphertext-only attack* and *message modification attack*, are related to data confidentiality and integrity. To protect the CE-CPS from these attacks, we introduce techniques to encrypt $\mathcal{QP}$, and design an efficient method to verify its solution from the cloud. The difference between the proposed encryption and other standard encryptions is that only the control system needs to encrypt and decrypt the data at the physical layer of the system.

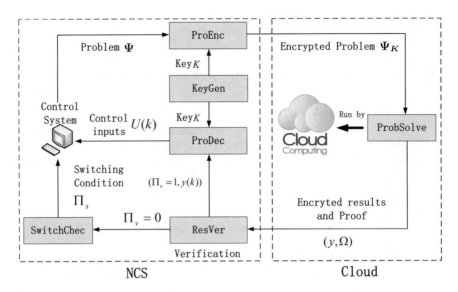

Fig. 4.2 The proposed CE-CPS mechanism interacts between CPS and the cloud

4.5.1 Encryption Methods

The first part of the mechanism is to encrypt the constraints of $\mathcal{QP}$. Let $G \in \mathbb{R}^{p \times p}$ and $F \in \mathbb{R}^{m \times p}$ be random non-singular matrices. We perform the following transformation,

$$\begin{cases} MU_k \leq c, \\ EU_k = d, \end{cases} \Rightarrow \begin{cases} (M + FE)U_k \leq c + Fd, \\ G(EU_k) = Gd. \end{cases}$$

Secondly, we encrypt the decision variable $U(k)$. Let $P \in \mathbb{R}^{lN \times lN}$ be a random non-singular matrix, i.e., P^{-1} exists, and $g \in \mathbb{R}^{lN \times 1}$ be a random non-zero vector. Then, we transform U_k into y_k by using a one-to-one mapping $y_k := P^{-1}(U_k - g)$, and y_k has the same size as U_k.

Using the key $K \triangleq (G, F, P, g)$, $\mathcal{QP}$ can be transformed into the following encrypted problem $\mathcal{QP}_K$ by

$$\mathcal{QP}_K : \min_{y(k)} \quad J_K(y_k) = y_k^T \tilde{Q} y(k) + \tilde{b}^T y_k$$

$$\text{s.t.} \quad \tilde{M} y_k \leq \tilde{c}, \ \tilde{E} y_k = \tilde{d},$$

where $\tilde{Q}$, $\tilde{M}$, $\tilde{E}$, $\tilde{b}$, $\tilde{c}$, and $\tilde{d}$ are matrices and vectors of appropriate dimensions given as follows:

$$\tilde{Q} := P^T Q P, \quad \tilde{M} := (M + FE)P, \quad \tilde{E} := GEP, \tilde{b} := (2g^T Q P + b^T P)^T,$$

$$\tilde{c} := c + Fd - (M + FE)g, \quad \tilde{d} := G(d - Eg). \tag{4.4}$$

We define $\mathcal{QP}_K$ by the tuple $\Psi_K := (\tilde{Q}, \tilde{M}, \tilde{E}, \tilde{b}, \tilde{c}, \tilde{d})$, which has the same size as Ψ. The following theorem characterizes the correctness of the encryption.

Theorem 4.1 *If y is the solution for* $\mathcal{QP}_K$, *then* $U_k = Py_k + g$ *must be the solution.*

Remark 4.1 Theorem 4.1 shows that the customized cryptography will not affect the correctness of the solutions.

4.5.2 Verification Methods

Due to the *message modification attack*, we need to verify the solution y from the cloud to achieve data integrity. To reduce the complexity of the verification, we find an easily verifiable sufficient and necessary condition for the solution.

Let $\mathcal{D}_K$ be the dual problem of $\mathcal{QP}_K$, and $d_K(\lambda, \nu)$ be the value of $\mathcal{D}_K$, where λ and ν are the dual decision variables. Since the Slater's condition is satisfied, so the strong duality of $\mathcal{QP}_K$ holds, i.e., $d_K(\lambda, \nu) = J_K(y)$, when λ and ν are solutions to $\mathcal{D}_K$, and y is a solution to $\mathcal{QP}_K$ [21]. Then, the cloud can solve $\mathcal{QP}_K$ using the Karush-Kuhn-Tucker (KKT) condition [21], which states that y is a solution to $\mathcal{QP}_K$ if and only if y, λ, and ν satisfy

$$\tilde{E}y - \tilde{d} = 0, \quad \tilde{M}y - \tilde{c} \leq 0, \tag{4.5}$$

$$\lambda \geq 0, \quad \lambda^T(\tilde{M}y - \tilde{c}) = 0, \tag{4.6}$$

$$\frac{\partial J_K}{\partial y} = 2\tilde{Q}y + \tilde{b} + \tilde{M}^T\lambda + \tilde{E}^T\nu = 0. \tag{4.7}$$

Based on the above results, we design a proof $\Omega := (\lambda, \nu)$ to be sent to the control system together with the solution y. The control system can verify y using (4.5)–(4.7).

A direct verification using (4.7) can be inefficient when N is large, because the computation required to check a $lN \times 1$ vector $\partial J_K/\partial y$ can be costly. To improve the efficiency, we can randomly check the elements in $\partial J_K/\partial y$. To this end, we define an $lN \times 1$ random vector $r = \{r_1, \ldots, r_{lN}\}$, where $r_i \sim \text{Bernoulli}(1/2)$, i.e.,

$$\Pr[r_i = 0] = \Pr[r_i = 1] = \frac{1}{2}, \text{ for } i = 1, \ldots, lN.$$

Using (4.7), we arrive at

$$r^T\left(2\tilde{Q}y + \tilde{b} + \tilde{M}^T\lambda + \tilde{E}^T\nu\right) = 0. \tag{4.8}$$

Hence, instead of checking (4.7) directly, we can check (4.8), which is a scalar equality. This verification can be performed q times by generating the random vector r q times.

Theorem 4.2 *The proposed mechanism ensures that a false solution cannot succeed in the verification mechanism with a probability greater than* $(1/2)^q$.

Remark 4.2 Theorem 4.2 indicates that we can achieve different security levels by choosing the parameter q. In general, we choose q much smaller than lN so that the verification can be more efficient than directly checking (4.7). However, a small q will lead to a low-security level. Therefore, there is a fundamental tradeoff between efficiency and security level.

4.6   Availability Issues

The third attack, *availability attack*, highlights the cyber-physical nature of CE-CPSs and can cause a serious problem for the stability of CE-CPSs as the control system requires feedback control inputs at every sampling time to stabilize itself. It can be easily implemented by an attacker who continuously sends erroneous control solutions from the cloud to the CPS. This time-critical property makes CPSs vulnerable to this type of threat.

To this end, we design a Switching Mode Mechanism (SMM) as part of our control design to guarantee the stability of CPSs and enhance their resilience to this attack even when no control inputs are available from the cloud either in a short or long duration. In addition, SMM allows the CPS to maintain a certain level of control performance, and recover on-line when an attacker succeeds in the attack.

4.6.1   Switching Mode Mechanism

In the SMM, we design three modes for the CPSs: Cloud Mode $\mathcal{N}_c$, Buffer Mode $\mathcal{N}_b$, and Safe Mode $\mathcal{N}_s$. Figure 4.3 illustrates how the SMM works for the CE-CPS. $\Pi_v \in \{0, 1\}$ is the output of the verification: $\Pi_v = 1$ means that the verification is successful, while $\Pi_v = 0$ means otherwise. $\Pi_s \in \{0, 1\}$ is the output of a switching condition, and $\Pi_s = 1$ means that a switching condition is satisfied, while $\Pi_s = 0$ means otherwise. Each working mode is described as follows:

1. **Cloud Mode** $\mathcal{N}_c$: In $\mathcal{N}_c$, the CPS uses a cloud-based controller, which outsources the computation of $\mathcal{QP}_K$ to the cloud. When $\Pi_v = 1$, the CPS stays in the $\mathcal{N}_c$.
2. **Buffer Mode** $\mathcal{N}_b$: In $\mathcal{N}_b$, the CPS uses the control inputs generated in the previous stage. For instance, if the CPS receives the last control input at time k, and no inputs are available from $k + 1$ to $k + \tau$, then at time $k + \tau$, the CPS uses control inputs $\hat{u}_{k+\tau|k}$ generated at time k when $\Pi_s = 0$.

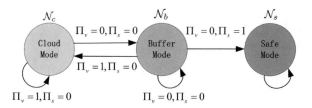

Fig. 4.3 The Switching Mode Mechanism: (1) Cloud Mode $\mathcal{N}_c$: if $\Pi_v = 1$, the CPS stays in this mode; if $\Pi_v = 0$, the CPS switches to Buffer mode. (2) Buffer Mode $\mathcal{N}_b$: if $\Pi_v = 1$ and $\Pi_s = 0$, then system switches back to the Cloud Mode. (3) Safe Mode $\mathcal{N}_s$: if $\Pi_s = 1$, then the CPS switches to the Safe Mode, and never switch back

3. **Safe Mode $\mathcal{N}_s$:** In $\mathcal{N}_s$, the CPS uses a local controller to stabilize itself. For example, if the CPS stays in Buffer Mode, and a switching condition is satisfied ($\Pi_s = 1$) at time $k + \tau$, indicating that the control inputs $\hat{u}_{k+\tau|k}$ cannot guarantee the stability for the CPS, then CPS switches to $\mathcal{N}_s$. This switching condition will be designed later in Sect. 4.2.

Here, we do not allow the CPS to switch back from the Safe Mode to the Cloud Mode. This is due to the fact that "switching back" may result in unanticipated oscillations, which can lead to a new threat.

4.6.2 Buffer Mode and Switching Condition

The design goal of the Buffer Mode is to guarantee the stability of the CPSs when the *availability attack* happens for a sufficiently short period of time. One simple way to tackle this problem is to use the remaining control inputs $\{\hat{u}_{k+1|k}, \ldots, \hat{u}_{k+\tau|k}\}$ generated at time k, when the attack occurs from $k+1$ to $k+\tau$. However, this solution is not ideal, since in real applications, disturbances and noises are ubiquitous in CPSs, and we cannot ensure that these control inputs can still stabilize the CPSs.

One sophisticated way to make use of the remaining control inputs is to apply an event-triggered MPC scheme [86], and find a switching condition, as illustrated in Fig. 4.2, that can stabilize the system through switching between $\mathcal{N}_c$ and $\mathcal{N}_b$. In addition, to capture the effect of disturbances, a modified model is given as follows:

$$x_{k+1} = Ax_k + Bu_k + Dw_k, \tag{4.9}$$

where $w_k \in \mathbb{R}^{n \times 1}$ is a disturbance vector bounded by $||w_k|| \leq \bar{w}$, and $D \in \mathcal{R}^{n \times n}$ is a constant matrix.

Based on (4.9) and the event-triggered MPC scheme, the following theorem presents a switching condition for the SMM.

Theorem 4.3 *Let k be the last sampling time that the CPS receives a correct solution from the cloud. At the time $k + \tau$ ($1 \leq \tau \leq N - 1$), the CPS cannot be guaranteed stability using the input $\hat{u}_{k+\tau|k}$ if*

$$\Pi_s : \sum_{i=0}^{N-1} \|A^i D\|^2 \bar{\omega} > \sigma h(x_{k+\tau-1}, \hat{u}_{k+\tau-1|k}), \tag{4.10}$$

where $h(x, u) = \|x\|^2 + \eta\|u\|^2$, and $\sigma \in (0, 1)$, is a tuning parameter. Otherwise, $\hat{u}_{k+\tau|k}$ can be used to stabilize the system, when no inputs are available from the cloud.

Remark 4.3 In Theorem 4.3, the idea of proving the stability of a control system is to use Lyapunov theory to demonstrate that the system energy decreases in time k when using the designed control inputs. To describe the system energy, we need to find an appropriate Lyapunov function. The details are given in the proof.

Remark 4.4 Since A, D, $\bar{\omega}$, and σ are given, the CPS only needs to compute $h(x_{k+\tau-1|k}, \hat{u}_{k+\tau-1|k})$ to check (4.10). Therefore, the computations of checking this switching condition is not heavy for the control system.

The algorithm **SwitchChec** is used to check the switching condition (4.10) and decrypt the results from the cloud. In this algorithm, the time-consuming part is the matrix-vector multiplication in the decryption and the computation of switching conditions. Therefore, the time complexity of this algorithm is $O(n^2)$.

4.6.3   The Local Controller for the Safe Mode

If (4.10) holds, the CPS switches to the Safe Mode $\mathcal{N}_s$ to use a local controller. Since the CPSs have no other information in $\mathcal{N}_s$, we use a perfect-state-feedback H_∞-optimal controller, which deals with the worst-case disturbance, through an offline design.

Let $k + \tau$ be the time that the switching condition (4.10) holds ($\Pi_s = 1$), and the CPS switches to $\mathcal{N}_s$, where k is the last sampling time that the CPS receives a correct y from the cloud. We denote $x_s(\hat{k})$, $u_s(\hat{k})$ and $w_s(\hat{k})$ as the state values, the control inputs, and the disturbances in $\mathcal{N}_s$, respectively, where $\hat{k} \geq k + \tau$. The system model in $\mathcal{N}_s$ is given by

$$x_s(\hat{k} + 1) = Ax_s(\hat{k}) + Bu_s(\hat{k}) + Dw_s(\hat{k}),$$
$$z_s(\hat{k}) = Cx_s(\hat{k}),$$

where $z_s \in \mathbb{R}^n$ is the system output, $C = I \in \mathbb{R}^{n \times n}$ is an identity matrix, and the initial condition of the state is $x_s(k + \tau) = x(k + \tau)$, which is the last state in $\mathcal{N}_b$.

The goal of the H_∞-optimal controller is to find an optimal control policy $u_s^*(\hat{k}) = \mu(x(\hat{k}))$ such that $\sup_{w_s} \|z_s\|/\|w_s\| \leq \gamma$, where $\gamma > 0$ is a given constant.

We assume that the pair (A, B) is stabilizable. As stated in [11], with a given parameter $\gamma > 0$, the H_∞-optimal controller for the infinite-horizon case is given by

$$u_s^*(\hat{k}) = \mu^*(x_s(\hat{k})) = K_s x_s(\hat{k}), \tag{4.11}$$

where

$$K_s := -B^T \Lambda (I + (BB^T - \gamma^{-2} DD^T)\Lambda)^{-1} A,$$

$$\Lambda := Q_x + A^T \Lambda (I + (BB^T - \gamma^{-2} DD^T)\Lambda)^{-1}.$$

The existence condition of the controller (4.11) is given by $\gamma^2 I - D^T \Lambda D > 0$. If we can find the smallest $\bar{\gamma}$ satisfying the above inequality, then for all $\gamma \geq \bar{\gamma}$, the CPS can be guaranteed bounded-input bounded-state (BIBS) stable [11].

The control gain K_s is computed off-line based on the knowledge of A, B, D, and γ. Therefore, the CPSs only need to compute (4.11) in $\mathcal{N}_s$. The H_∞-optimal controller is more costly in comparison to MPC because it deals with the worst-case disturbance w_s^*, which leads to more conservative and expensive control inputs. Hence, the H_∞ controller is suitable for the Safe Mode $\mathcal{N}_s$, when the cloud is not available.

4.7 Analysis and Experiments

A UAV example is used in the experiments. The computations of the UAV and the cloud are performed on the different blocks in the Simulink of MATLAB 2014b, running on the workstation with an Intel Core i7-4770 processor and a 12-G RAM.

The UAV is a small-scale unmanned helicopter, whose dynamic model is linearized at its equilibrium point (hovering point), and the state x of the UAV is defined as

$$x = \left[p_x, \ p_y, \ p_z, \ v_x, \ v_y, \ v_z, \ \phi, \ \theta, \ \psi, \ p_a, \ q_a, \ r_a, \ a_s, \ b_s \right],$$

where (p_x, p_y, p_z) are the position on (X, Y, Z)-axis of the north-east-down (NED) frame, (v_x, v_y, v_z) are the velocity on (X, Y, Z)-axis of the body frame, (ϕ, θ, ψ) are Euler angles, (p_a, q_a, r_a) are roll, pitch, and yaw angular rates, (a_s, b_s) are the longitudinal and lateral flapping angle of the main rotor. The elements of matrices A and B in (4.1) are given in [22].

Table 4.1 Efficiency tests of the proposed mechanism

Benchmark		Original prob.	Crypto over.	Encryption	Cloud efficiency
#	Length	t_{orig} (s)	t_{cry} (s)	t_{cloud} (s)	t_{orig}/t_{cloud} (s)
1	$N = 10$	0.006757	0.000159	0.007001	0.965153
2	$N = 50$	0.147428	0.003251	0.164543	0.895984
3	$N = 100$	0.254544	0.008299	0.269742	0.943657
4	$N = 300$	0.634718	0.014426	0.768352	0.826088
5	$N = 500$	1.066516	0.028876	1.609127	0.664027

1. **Efficiency Tests:** The length of the horizon window N determines the size of matrix Q, which is related to the complexity of $\mathcal{QP}$, and hence, we increase N from 10 to 500 to observe the relationship between N and computational time. The verification parameter q is 20. Table 4.1 summarizes the experimental results, where each entry in the table is the average computational time of 10 identical feedback control experiments. In Table 4.1, the computational time to solve $\mathcal{QP}$, t_{orig}, is given in the third column. The time to solve $\mathcal{QP}_K$, t_{cloud}, is given in the fifth column. The time of the cryptographic overhead, t_{cry}, including the key generation, encryption, decryption, and verification, is given in the fourth column. From the results, t_{cry} is mush less than t_{orig}; i.e., outsourcing computations to a cloud can enhance efficiency of the UAV. In addition, the cloud efficiency, calculated as t_{orig}/t_{cloud}, is given in the seventh column. Normally, the encryption should not increase the time to solve the quadratic problem. Thus, the cloud efficiency should be close to 1.

2. **Stability and Resiliency Tests:** The next numerical experiment is used to test the stability and resiliency of the CE-CPS in the *message modification attack* and *availability attack*. The control objective of the UAV is to track an ascending helix trajectory under a disturbance w_k bounded by 0.5. We set the parameter σ in (4.10) as 0.8. We first show that when no attack occurs, the UAV succeeds in tracking the desired trajectory, and the result is illustrated in Fig. 4.4, where the solid red curve is the reference trajectory, and the blue dash curve is the real trajectory. Then, we conduct other experiments in four cases to verify our mechanism:

   (a) No Verification: verification algorithms in Sect. 3.2 are ignored for the solutions. The UAV receives a wrong solution from the cloud at time $k = 60$.
   (b) *Availability attack* of a short duration without SMM: the control inputs from $k = 60$ to $k = 65$ are not available due to the attack. The UAV without SMM have no control inputs, i.e., $u_k = 0$, in this duration.
   (c) *Availability attack* of a short duration with SMM: the control inputs from $k = 60$ to $k = 65$ are unavailable due to the attack.
   (d) *Availability attack* of long duration with SMM: the control inputs are unavailable from $k = 60$ to the end of the experiment.

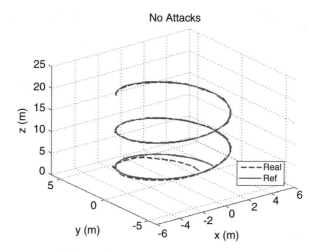

Fig. 4.4 No Attack: The UAV tracks the trajectory smoothly when no attack occurs

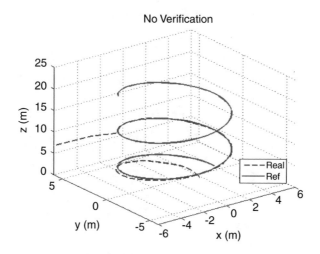

Fig. 4.5 Case 1: no verification for the solutions

Figures 4.5, 4.6, 4.7, 4.8 illustrate experimental results for the four cases. In Fig. 4.5, we can see that without verification, one wrong solution from the cloud deviates the UAV from the reference trajectory. Figures 4.7 and 4.8 show that the proposed SMM mechanism guarantees that the UAV tracks reference trajectory successfully, despite the unavailability of the control inputs from the cloud for either a long or a short duration.

To better understand when the UAV switches from the cloud mode to other modes and the effect of σ on the switching condition (4.10), we present the tracking errors in Figs. 4.9 and 4.10 under different scenarios. Figure 4.9 illustrates the tracking errors in a short-duration attack with $\sigma = 0.8$. The UAV switches to the buffer

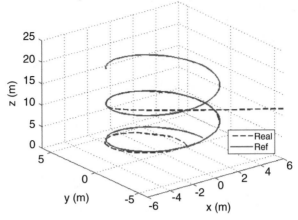

Fig. 4.6 Case 2: availability attack without SMM

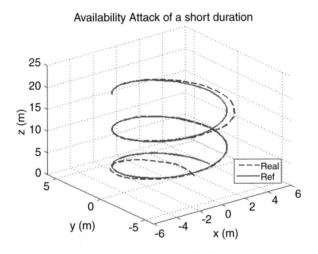

Fig. 4.7 Case 3: availability attack of a short duration with SMM

mode $\mathcal{N}_b$ at time $k = 60$, and switches back to the cloud mode $\mathcal{N}_c$ at time $k = 66$. Figure 4.10 shows the tracking errors in long-duration attacks with $\sigma = 0.6$ and 0.8. When $\sigma = 0.6$, the UAV switches to the buffer mode $\mathcal{N}_b$ at time $k = 60$, and switches to the safe mode $\mathcal{N}_s$ at time $k = 68$. When $\sigma = 0.8$, the UAV switches to the buffer mode $\mathcal{N}_b$ at time $k = 60$, and switches to the safe mode $\mathcal{N}_s$ at time $k = 70$. It is clear that the smaller the σ is, the more quickly the system switches to the safe mode, allowing the system to recover faster. However, when σ is small, the switching mechanism is more sensitive to disturbance ω, i.e., the system may switch to the safe mode inaccurately due to a large disturbance while no attack occurs.

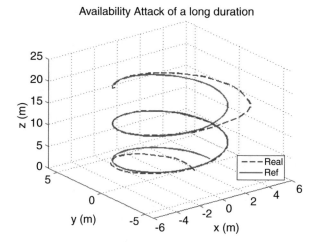

Fig. 4.8 Case 4: availability attack of a long duration with SMM

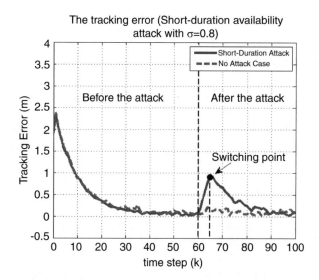

Fig. 4.9 The tracking errors in a short-duration attack with $\sigma = 0.8$

The experimental results above highlight that traditional cryptographic tools are not sufficient to meet the CE-CPS security requirements due to its cyber-physical nature. Traditional encryption and verification can protect data confidentiality and integrity, but they fail to tackle the stability issue of CE-CPSs. Therefore, we incorporate control tools into our mechanism to solve this problem. Our results demonstrate that the proposed mechanism can not only achieve data confidentiality and integrity but also guarantee stability and enhance the CE-CPS resiliency.

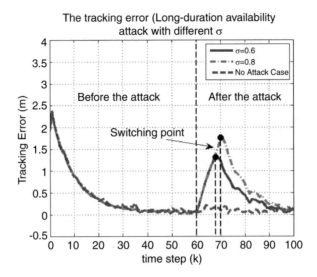

Fig. 4.10 A long-duration attack with σ equals to 0.8 and 0.6, respectively

4.8   Conclusions and Notes

In this chapter, we have designed security and resiliency mechanisms for CE-CPSs with the presence of adversaries. We have formulated a cloud-enabled model predictive control problem for CPSs and presented three attack models that have an impact on the integrity, confidentiality, and availability of the CPS. To achieve confidentiality and integrity, we have developed efficient methods to encrypt the problem and verify the solution from the cloud. To protect the system from the availability attack, we have designed a switching mechanism to ensure the stability and resiliency despite the unavailability of control inputs from the cloud. We have seen from the case study of the UAVs that the customized encryption technique has achieved desirable security performance. The systems can recover from the availability attack from both short and long duration of attacks.

We have introduced customized cryptographic techniques that are protect the data confidentiality and integrity in control systems. The encryption techniques introduced in this chapter can be further extended to other control techniques. For example, in [108], similar techniques have been applied to a class of linear systems against stealthy data injection attacks. One issue that has not been fully discussed is the security guarantee of the customized encryption. If the key does not change over a long period of time, an attacker can use brute-force attacks to obtain the okey. Hence, there is a need to build a mechanism to change keys strategically on regular basis. The frequency of updating the keys relies on the attacker's capabilities and the criticality of the mission. This problem can be addressed by FlipIt game introduced in Chap. 6. Interested can also refer to an application of FlipIt game in Chap. 7.

The resiliency mechanisms are pivotal to CPS. In this chapter, we have introduced a switching mechanism to enable a fast response to degradation in performance. The controller design relies on techniques from model predictive control [23, 102], event-triggered control [43, 58], and H_∞ optimal control theory [11, 51]. We have presented a discrete-time linear-quadratic regulator (LQR) problem. For background on basics of optimization, readers can refer to Appendix A. Readers can refer to Appendix B for a succinct background on LQR optimal control theory. Resilient control systems, as have been discussed in [137] and [141], are a class of control systems capable of self-healing and self-reconfigurability. The conceptual frameworks have been introduced in [140, 142, 199]. The common designs of such systems rely on automated switching between two control mechanisms, as in [25, 31, 182] and the modeling of adversarial behaviors as in [201, 202]. Interested readers can refer to [65] for a recent review on systems and control perspective of CPS security and [40] for a recent review on secure and resilient control system design using dynamic game theory.

Chapter 5
Secure Data Assimilation of Cloud Sensor Networks

5.1 Introduction to CE-LSNs

Large-scale Sensing Networks (LSNs), such as power grids [152], pollution sensing system, and transportation system, often require extensive sensing information from remote sensors. Given the measurements, the fusion center of the LSNs needs to reconstruct the entire state information, known as data assimilation. The performance (e.g., efficiency and accuracy) of the data assimilation is critical to the LSNs. However, with the increasing size of the sensing data, new challenges arise in LSNs. Firstly, with an increasing the number of remote sensors, it is challenging for the fusion center to assemble all the data due to the limited bandwidth of the networks. Secondly, the complexity of the data assimilation problem grows with the dimensions of the state of the system. The increasing number of sensors will make the computation cost impractical for the fusion center in the future.

The advent of Cloud Computing Technologies (CCTs) makes it possible to tackle the computational issue. The integration of CCTs with control and sensing systems brings revolutionary features, such as massive computation resources and real-time data processing [8]. The integration of LSNs with CCTs, resulting in cloud-enabled LSNs, can also bring substantial benefits to the systems, especially by enhancing the efficiency of large-scale data assimilation. Figure 5.1 illustrates the new architecture of the CE-LSNs, which aims to improve the effectiveness of the computations in LSNs. On the one hand, all the sensors can use their local networks to upload the data to the cloud, instead of directly sending to the fusion center, whose network has limited bandwidth. On the other hand, with massive computation resources, the cloud can solve the data assimilation problem efficiently.

Despite the advantages of CE-LSNs, the cyber-physical security issues arise due to the untrusted cloud. Directly sending the sensing information to the cloud may expose the privacy of sensitive information. Adversaries can steal valuable

© The Editor(s) (if applicable) and The Author(s), under exclusive license to
Springer Nature Switzerland AG 2020
Q. Zhu, Z. Xu, *Cross-Layer Design for Secure and Resilient Cyber-Physical Systems*,
Advances in Information Security 81, https://doi.org/10.1007/978-3-030-60251-2_5

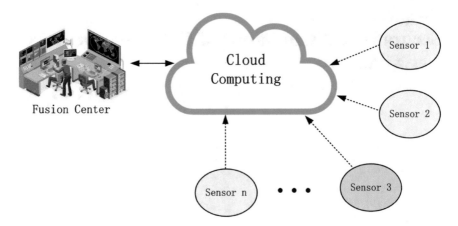

Fig. 5.1 The large-scale sensing networks with the aid of cloud computing: each sensor uploads its sensing information to a cloud; the fusion center outsources the computation to the cloud based on the sensing information in the cloud; finally, the cloud return the results to the fusion center

information by hacking into the cloud or eavesdropping the communications between the cloud and sensors [198]. One possible approach is to apply the homomorphic encryption to encrypt data before sending it to the cloud. The ideal homomorphic encryption allows the cloud to process operations, such as addition and multiplication, on the ciphertexts, generating an encrypted results, which can be decrypted to the desired results. The ideal homomorphic encryption means that the encryption can achieve homomorphic property both in multiplication and addition. However, this ideal encryption is hard to achieve (see [85]). Partially homomorphic encryptions are easy to realize; i.e., the encryption can only achieve homomorphic property either in multiplication or addition. For example, the RSA and *ElGamal* can realize homomorphic multiplication [80]; i.e., the cloud can process the multiplication on the ciphertexts, while Benaloh and Paillier cryptosystems can achieve homomorphic addition.

In this chapter, we aim to design a secure mechanism for LSNs to outsource the data assimilation to an untrusted cloud securely. Since the fully homomorphic encryption is inefficient to realize, we combine the traditional and customized encryption scheme to achieve fully homomorphic property. The main issue of the customized encryption is the requirement of changing the key for each mission. The high frequency in changing the key may introduce new challenges for the system. Hence, in our mechanism, only the fusion center applies the customized encryption, while the sensors use the standard encryption. We present the correctness of the and present analysis of the security and efficiency of the mechanism. Besides, the encryption introduces quantization errors in the estimation problem. We analyze the impact on the performance of the estimation problem.

5.2 Problem Formulation

In this section, we first use a linear system to describe the dynamics of large-scale sensing networks. Then, we design a Kalman filter for data assimilation. Due to the large-scale property of LSNs, we aim to outsource the computation of data assimilation to a cloud to increase efficiency. However, outsourcing the computation to an untrusted cloud introduce security challenges. Hence, we propose the design objectives of our secure mechanism at the end of this section.

5.2.1 System Model and the Outsourcing Kalman Filter

The discrete-time model of a large-scale LSN is given by

$$x_{k+1} = A_k x_k + w_k, \quad y_k = C_k x_k + v_k,$$

where $x_k \in \mathbb{R}^n$ is the state of the LSNs with a given initial condition $x_0 \in \mathbb{R}^{n_x}$; $y_k \in \mathbb{R}^{n_y}$ is the sensing information, and $n_y (n_y \leq n_x)$, is the number of the sensors; matrix $A_k \in \mathbb{R}^{n_x \times n_x}$ describe the dynamic movement of the state x; matrix $C_k \in \mathbb{R}^{n_y \times n_x}$ describe the topology of the sensors; $w_k \in \mathbb{R}^{n_x}$ and $v_k \in \mathbb{R}^{n_y}$ are the additive Gaussian noise with zero means and covariances, Σ_w and Σ_v, respectively.

In an LSN, each sensor $i, i = 1, \ldots, m$, uploads its sensing data $y_i(k)$ to a fusion center. The fusion center processes data assimilation to estimate x_k based on the feedback information y_k. Hence, we can design a Kalman filter to achieve this goal. The Kalman filter is given by

$$\hat{x}_{k+1} = A_k \hat{x}_k + H_k(y_k - C_k \hat{x}_k) = A_k \hat{x}_k + H_k(y_k - z_k) = A_k \hat{x}_k + \phi_k, \quad (5.1)$$

where $\hat{x}_k \in \mathbb{R}^n$ is the estimated state of x_k; $z_k := C_k \hat{x}_k \in \mathbb{R}^{n_y}$ is the estimated output; $\phi_k := H_k(y_k - z_k) \in \mathbb{R}^{n_x}$ is the regularized term. The explicit forms of the matrices $H_k \in \mathbb{R}^{n_x \times n_y}$ and $P_k \in \mathbb{R}^{n_x \times n_x}$ are given by

$$H_k = A_k P_k C_k'(\Sigma_v + C_k P_k C_k')^{-1}, \quad (5.2)$$

$$P_{k+1} = \Sigma_w + (A_k - H_k C_k) P_k A_k'. \quad (5.3)$$

Hence, based on the sensing information y_k from each sensor, the fusion center can estimate the state x_k by computing (5.1)–(5.3), iteratively.

5.2.2   Challenges and Design Objectives

In an LSN, the number of the sensors n_y is large. The direct computation of (5.1)–(5.3) brings heavy computation for the fusion center, which may not be equipped with powerful computational resources. Therefore, with the development of the cloud computing technology, outsourcing the computation (5.1)–(5.3) to a cloud can increase efficiency and reduce the computation burden for the fusion center. Besides increasing efficiency, one of significant advantages in cloud computing is data sharing. Instead of sending the information to the fusion center, all the sensors in the system can directly upload their sensing information to the cloud for data assimilation. This approach significantly reduces the computation and transmission cost of the fusion center.

Despite the advantages, the integration with an untrusted cloud inevitably introduces new security challenges to the system. An adversary can access the cloud can observe the sensing information y_k and the system information A_k and C_k stored temporally in the cloud. It motivates us to develop a new mechanism to secure outsourcing the computation of data assimilation. The design objectives are summarized as follows:

- Data Confidentiality: The secure mechanism should protect the privacy of the data sharing in the cloud, including the sensing information y_k, the estimated output z_k, and the system matrices A_k and C_k.
- Efficiency: The proposed scheme should reduce the computation burden of the fusion center, and should not add too much load for the sensors. The sensor only needs to process the encryption algorithm, while the fusion center only processes encryption and decryption algorithms without other computations.

Figure 5.2 illustrates the architecture of the proposed secure mechanism. Firstly, each sensor in the system encrypts its sensing information $y_i(k)$ and sends the ciphertext $\bar{y}_i(k)$ to the cloud. Simultaneously, the fusion sends the encrypted

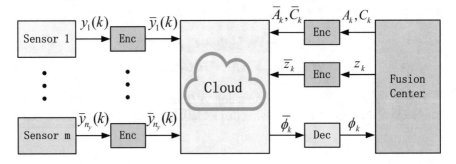

Fig. 5.2 The large-scale sensing networks with the aid of cloud computing: each sensor uploads its sensing information to a cloud; the fusion center outsources the computation to the cloud based on the sensing information in the cloud; finally, the cloud returns the results to the fusion center

information $\bar{A}_k$, $\bar{C}_k$, and $\bar{z}_k$ to the cloud. The cloud performs the computation on the ciphertexts and returns the encrypted results $\boldsymbol{\phi}_k$ to the fusion center. After decryption, the fusion center obtains the desired regularized term ϕ_k in (5.1) to update the estimation.

5.3 The Secure Outsourcing Data Assimilation

To achieve the designed objectives given in Sect. 5.2, we need to use a homomorphic encryption scheme. The basic idea of homomorphic encryption is to allow a third party (e.g., a cloud) to process computations on the ciphertexts, generating encrypted results. The encrypted results, when decrypted, matches the results of the operations performing on the original information. However, a fully homomorphic encryption scheme, which can achieve both homomorphic addition and multiplication, is hard to reach. To this end, we need to combine conventional and customized encryption methods.

5.3.1 The Additive Homomorphic Encryption

Even the fully homomorphic encryption is challenging to design; the additively homomorphic encryption is achievable. The following definition presents the property of the additively homomorphic encryption.

Definition 5.1 An encryption method $(\mathcal{E}, \mathcal{D}, PK, SK)$ is said to be additively homomorphic if it satisfies

$$
\mathcal{D}_{SK}\left[\bar{m}_1 \cdot \bar{m}_2\right] = \mathcal{D}_{SK}\left[\mathcal{E}_{PK}(m_1) \cdot \mathcal{E}_{PK}(m_1)\right]
$$

$$
= \mathcal{D}_{SK}\left[\mathcal{E}_{PK}(m_1)\right] + \mathcal{D}_{SK}\left[\mathcal{E}_{PK}(m_2)\right] = m_1 + m_2,
$$

$$
\mathcal{D}_{SK}\left[\bar{m}_1^l\right] = l \cdot \mathcal{D}_{SK}\left[\mathcal{E}_{PK}(m_1)\right] = l \cdot m_1,
$$

where PK and SK are the public key and secret key, respectively; $\mathcal{E}_{PK}(\cdot)$ and $\mathcal{D}_{SK}(\cdot)$ are the encryption and decryption algorithms, respectively; $\bar{m}_1, \bar{m}_2 \in \mathcal{C}$ are the ciphertexts; $m_1, m_2 \in \mathcal{M}$ are the plaintexts; $l \in \mathcal{M}$ is a constant. $\mathcal{M}$ and $\mathcal{C}$ are the sets of the plaintexts and ciphertexts, respectively.

Remark 5.1 The additively homomorphic encryption allows us to process additive operations to the ciphertexts without decryption, but it also shows that it cannot conceal the parameter $l \in \mathcal{M}$ when we handle multiplication.

5.3.2 The Homomorphic Observer

Based on the homomorphic encryption, we aim to design a homomorphic observer that allows the fusion center securely outsource the computations of (5.1)–(5.3) to an untrusted cloud. To this end, the following statement defines a homomorphic observer.

Definition 5.2 Define a mapping $\mu : \mathbb{R}^{n \times m} \times \mathbb{R}^m \times \mathbb{R}^m \to \mathbb{R}^n$ such that $\phi_k = \mu(H_k, y_k, z_k,) := H_k(y_k - z_k)$. Given an encryption scheme $(\mathcal{E}, \mathcal{D}, PK, SK)$, if a mapping $\mu_{\mathcal{E}} : \mathbb{R}^{n \times m} \times \mathcal{C}^m \times \mathcal{C}^m \to \mathcal{C}^n$ satisfies

$$\mu_{\mathcal{E}}\left(H_k, \bar{y}_k, \bar{z}_k \right) = \mathcal{E}_{PK}\left[\mu(H_k, y_k, z_k) \right] = \bar{\phi}_k, \qquad (5.4)$$

where $\bar{y}_k = \mathcal{E}_{PK}(y_k)$, $\bar{z}_k = \mathcal{E}_{PK}(z_k)$, and $\bar{phi}_k = \mathcal{E}_{PK}(\phi_k)$, then we say $\mu_{\mathcal{E}}(\cdot)$ is a homomorphic observer of $\mu(\cdot)$.

Remark 5.2 According to Definition 5.2, if we can find a mapping $\mu_{\mathcal{E}}$ satisfying (5.4), then, the sensors and the fusion center can send the encrypted message $\bar{y}_k$ and $\bar{z}_k$ to the cloud, respectively. The cloud can run the computation of $\mu_{\mathcal{E}}(H_k, \bar{y}_k, \bar{z}_k)$ and send encrypted results $\bar{\phi}_k$ back to the fusion center. This approach protects the privacy of the sensing information y_k and the estimated output z_k.

Now, we aim to design a homomorphic observer based on the Paillier cryptosystem that satisfies the additively homomorphic properties given in Definition 5.1. The following theorem presents the explicit form of the desired homomorphic observer $\mu_{\mathcal{E}}$.

Theorem 5.1 *Suppose that a homomorphic encryption scheme $(\mathcal{E}, \mathcal{D}, PK, SK)$ satisfies the properties in Definition 5.1. Assume that the quantization errors approximately equal to zero. We define a mapping $\mu_{\mathcal{E}}(\cdot)$ as*

$$\mu_{\mathcal{E}}\left(H_k, \bar{y}_k, \bar{z}_k \right) := \begin{bmatrix} \mu_{\mathcal{E},1}\left(\{h_{1j}\}_{j=1}^m, \bar{y}_k, \bar{z}_k \right) \\ \vdots \\ \mu_{\mathcal{E},n}\left(\{h_{nj}\}_{j=1}^m, \bar{y}_k, \bar{z}_k \right) \end{bmatrix},$$

$$\mu_{\mathcal{E},i}\left(\{h_{1j}\}_{j=1}^m, \bar{y}_k, \bar{z}_k \right) := \prod_{j=1}^m \left(\bar{y}_j(k) \Big/ \bar{z}_j(k) \right)^{h_{ij}(k)},$$

where $\mu_{\mathcal{E},i}(\cdot)$ is the i-th element of the vector $\mu_{\mathcal{E}}(\cdot)$. Then, $\mu_{\mathcal{E}}(\cdot)$ is a homomorphic mapping of $\mu(\cdot)$, i.e.,

$$\mu_{\mathcal{E},i}\left(\{h_{1j}\}_{j=1}^m, \bar{y}_k, \bar{z}_k\right) = \mathcal{E}_{PK}\left[\mu_i\left(\{h_{1j}\}_{j=1}^m, y_k, z_k\right)\right],$$

where $\mu_i(\cdot)$ is the i-th element of the vector $\mu(\cdot)$.

Proof Let $\phi_i(k)$ be the i-th element of the vector ϕ_k. Given (5.1), we have

$$\phi_i(k) = \sum_{j=1}^m h_{ij}(k)y_j(k) - \sum_{j=1}^m h_{ij}(k)z_j(k).$$

Define that $\boldsymbol{\phi}_i(k) := \mu_{\mathcal{E},i}(\{h_{ij}\}_{j=1}^m, \bar{y}_k, \bar{z}_k)$. To complete the proof, we need to show that $\mathcal{D}_{SK}\left[\boldsymbol{\phi}_i(k)\right] = \phi_i(k)$. We observe that

$$\mathcal{D}_{SK}\left[\boldsymbol{\phi}_i(k)\right] = \mathcal{D}_{SK}\left[\prod_{j=1}^m (\bar{y}_j(k))^{h_{ij}(k)} \prod_{j=1}^m (\bar{z}_j(k))^{-h_{ij}(k)}\right]$$

$$= \mathcal{D}_{SK}\left[\prod_{j=1}^m \left(\bar{y}_j(k)\right)^{h_{ij}(k)}\right] + \mathcal{D}_{SK}\left[\prod_{j=1}^m \left(\bar{z}_j(k)\right)^{-h_{ij}(k)}\right]$$

$$= \sum_{j=1}^m \mathcal{D}_{SK}\left[\left(\bar{y}_j(k)\right)^{h_{ij}(k)}\right] + \sum_{j=1}^m \mathcal{D}_{SK}\left[\left(\bar{z}_j(k)\right)^{-h_{ij}(k)}\right]$$

$$= \sum_{j=1}^m h_{ij}(k)y_j(k) - \sum_{j=1}^m h_{ij}(k)z_j(k) = \phi_i(k).$$

As a result, $\mu_{\mathcal{E}}(\cdot)$ is a homomorphic mapping of $\mu(\cdot)$. $\square$

Remark 5.3 Given Theorem 5.1, the cloud can run the function $\mu_{\mathcal{E}}(\cdot)$ based on the ciphertexts $\bar{y}_k$ and $\bar{z}_k$. Hence, all the sensors can directly upload the ciphertexts $\bar{y}_k$ to the cloud for the data assimilation.

5.3.3   Customized Encryption for Outsourcing Computation

Theorem 5.1 presents that the cloud can process the computation of (5.1) with the ciphertexts $\bar{y}_k$ and $\bar{z}_k$, but it cannot provide protection for outsourcing the calculation (5.2) and (5.3). Note that the calculation of (5.2) and (5.3) includes both addition and multiplication. Hence, we need to design a customized encryption to achieve confidentiality for outsourcing (5.2) and (5.3) to the untrusted cloud.

The basic idea of our proposed encryption scheme is changing the coordinate of the state x_k. Let $(\mathcal{G}_C, \mathcal{E}_C, \mathcal{D}_C)$ be the customized encryption scheme, where $\mathcal{G}_C(\cdot)$,

$\mathcal{E}_C(\cdot)$, and $\mathcal{D}_C(\cdot)$ are the key generation, encryption, and decryption algorithms. The details of these three algorithms are presented below:

- $\mathcal{G}_C$: Given a random secret seed s, generate a random
- $\mathcal{E}_C$: On the input A_k, C_k, and P_k, the encryption algorithm yields that $\mathcal{E}_C(A_k, C_k, P_k) = (\bar{A}_k, \bar{C}_k, \bar{P}_k)$, where

$$\bar{A}_k := S_k A_k S_k', \quad \bar{C}_k := C_k S_k', \quad \bar{P}_k := S_k P_k S_k'. \tag{5.5}$$

- $\mathcal{D}_C$: On the input $\bar{H}_k \in \mathbb{R}^{n \times m}$ and $\bar{P}_{k+1} \in \mathbb{R}^{n \times n}$, $\mathcal{D}_C(\bar{H}_k, \bar{P}_{k+1}) = (S_k' \bar{H}_k, S_k' \bar{P}_{k+1} S_k)$, where

$$\bar{H}_k = \bar{A}_k \bar{P}_k \bar{C}_k' (\Sigma_v + \bar{C}_k \bar{P}_k \bar{C}_k')^{-1}, \tag{5.6}$$

$$\bar{P}_{k+1} = \bar{\Sigma}_w + (\bar{A}_k - \bar{H}_k \bar{C}_k) \bar{P}_k \bar{A}_k', \tag{5.7}$$

and $\bar{\Sigma}_w = S_k \bar{\Sigma} S_k'$.

The first thing that we need to show is the correctness of the customized encryption, which is characterized by the following theorem.

Theorem 5.2 *Given the customized encryption $(\mathcal{G}_C, \mathcal{E}_G, \mathcal{D}_C)$, the decrypted results $\mathcal{D}_C(\bar{H}_k, \bar{P}_{k+1})$ are identical to the results defined by (5.2) and (5.3).*

Proof Note that S_k is an orthogonal matrix, i.e., $S_k S_k' = I_n$, where $I_n \in \mathbb{R}^{n \times n}$ is an identity matrix. Substituting (5.5) into (5.6) and (5.7) yields that

$$\bar{H}_k = \bar{A}_k \bar{P}_k \bar{C}_k'(\Sigma_v + \bar{C}_k \bar{P}_k \bar{C}_k')^{-1}$$
$$= S_k A_k \underbrace{S_k' \bar{P}_k S_k}_{=P_k} C_k'(\Sigma_v + C_k \underbrace{S_k' \bar{P}_k S_k}_{=P_k} C_k')^{-1} = S_k H_k,$$

$$\bar{P}_{k+1} = \bar{\Sigma}_w + (\bar{A}_k - \bar{H}_k \bar{C}_k) \bar{P}_k \bar{A}_k'$$
$$= S_k \Sigma_w S_k' + S_k (A_k S_k' - H_k C_k) S_k' \bar{P}_k S_k A_k' S_k'$$
$$= S_k \left\{ \Sigma_w + (A_k - H_k C_k) P_k A_k' \right\} S_k' = S_k P_{k+1} S_k'.$$

Hence, it is easy to verify that

$$\mathcal{D}_C \begin{bmatrix} \bar{H}_k \\ \bar{P}_{k+1} \end{bmatrix} = \begin{bmatrix} S_k' \bar{H}_k \\ S_k \bar{P}_{k+1} S_k' \end{bmatrix} = \begin{bmatrix} H_k \\ P_{k+1} \end{bmatrix}.$$

This shows the correctness of the customized encryption scheme. □

Given the Paillier cryptosystem and the customized encryption scheme, we can allow the cloud to process the computation of the data assimilation based on the ciphertexts; i.e., the cloud can directly run the computations (5.6), (5.7), and

$$\tilde{\phi}_k = \mu_{\mathcal{E}}(\bar{H}_k, \bar{y}_k, \bar{z}_k).$$ (5.8)

However, since we encrypt the matrix H_k, directly decrypting $\tilde{\phi}_k$ does not give the solutions, i.e., $\mathcal{D}_{SK}(\tilde{\phi}_k) \neq \phi_k$. In the following theorem, we show how to obtain the solution ϕ_k from the ciphertexts $\tilde{\phi}_k$.

Theorem 5.3 *Given the homomorphic encryption scheme $(\mathcal{E}, \mathcal{D}, PK, SK)$ and the customized scheme $(\mathcal{G}_C, \mathcal{E}_C, \mathcal{D}_C)$,*

$$\phi_k = S'_k \cdot \mathcal{D}_{SK}(\tilde{\phi}_k),$$ (5.9)

where ϕ_k is defined by (5.8).

Proof Let $\tilde{\phi}_i(k)$ be the i-th element of the vector $\tilde{\phi}_k$. Given (5.8), we observe that

$$\mathcal{D}_{SK}[\tilde{\phi}_i(k)] = \mathcal{D}_{SK}[\mu_{\mathcal{E},i}(\{\bar{h}_{ij}(k)\}_{j=1}^m, \bar{y}_k, \bar{z}_k)]$$

$$= \mathcal{D}_{SK}\left[\prod_{j=1}^m (\bar{y}_j(k)/\bar{z}_j(k))^{\bar{h}_{ij}(k)}\right]$$

$$= \sum_{j=1}^m \bar{h}_{ij}(k)y_j(k) - \sum_{j=1}^m \bar{h}_{ij}(k)z_j(k).$$ (5.10)

Accordingly, the right hand side of (5.9) becomes

$$S'_k \cdot \mathcal{D}_{SK}[\tilde{\phi}_k] = S'_k \bar{H}_k(y_k - z_k) = S'_k S_k H_k(y_k - z_k) = \phi_k.$$ (5.11)

This completes the proof.   □

Theorem 5.3 indicates that the cloud can perform the computation of (5.1)–(5.3) based on the ciphertexts and return $\tilde{\phi}_k$ defined by (5.8) to the fusion center. Then, the fusion center can obtain ϕ_k using (5.9) and update the matrix P_{k+1} using the algorithm $\mathcal{D}_C(\bar{H}_k, \bar{P}_{k+1})$. In the next part, we will analyze the efficiency and security of the proposed mechanism.

5.4 Analysis of the Efficiency and Security

Now, we illustrate the proposed mechanism, combining the Paillier and customized cryptosystems, can achieve the designed objectives given in Sect. 5.3.

5.4.1   Efficiency Analysis

First of all, the cloud can increase the efficiency of assembling the data from the remote sensor. Secondly, cloud computing reduces computation complexity for the system. Specifically, the computation complexity of the Paillier cryptosystem is no greater than $O(n^2)$, and the calculation of the customized cryptosystem is the matrix multiplication, whose complexity is no greater than $O(n^3)$. However, the complexity of the Kalman filter is greater than $O(n^3)$. Hence, the proposed mechanism indeed enhance the performance of the data assimilation.

Another aspect related to efficiency is key management. Since all the sensors use the Paillier cryptosystem, it is easy for the fusion center to update the keys, i.e., the fusion can directly broadcast the public keys to all the sensors. For the customized encryption, only the fusion center requires the key to process the encryption $\mathcal{E}_C(\cdot)$ and decryption $\mathcal{D}_C(\cdot)$.

5.4.2   Security Analysis

In our proposed mechanism, we aim to protect the privacy of all the messages sent to the cloud, including the sensing information y_k, estimated output z_k, and system information A_k, C_k. Firstly, based on the Paillier cryptosystem, it is computationally difficult for an attacker to guess the plaintext y_k and z_k from the ciphertexts $\bar{y}_k$ and $\bar{z}_k$. Secondly, the customized cryptosystem changes its secret key S_k at each time k. Hence, observing the ciphertexts $\bar{A}_k$ and $\bar{C}_k$ cannot provide any information for the attackers about the system matrices A_k and C_k. Finally, the key management of the Paillier and customized schemes are completely independent. The independent key management ensures that even if attackers break one encryption scheme, they cannot make use of any information to break the other one. Therefore, our proposed mechanism provides two layers of protection to the outsourcing process.

5.5   Analysis of Quantization Errors

In the previous section, we assume that the quantization errors approximately equal to zero. In the encryption scheme, both the plaintexts and ciphertexts should be a positive integer, but normally, the sensing information is a real number. Hence, the encryption inevitably introduces quantization errors. In this section, we will analyze how the quantization errors affect the accuracy of the data assimilation.

Note that in the Paillier cryptosystem, the message should m should be a positive integer. Hence, we need to design a quantizer to quantize the state x_k. Define that $\tilde{x}_i(k) = q(x_i(k)) := \lceil \eta \cdot x_i(k) \rfloor$, where the scaling factor $\eta > 0$ is a sufficiently large integer, and $\lceil \cdot \rfloor$ is the rounding function. To obtain the plaintext m_k^y and m_k^z, we define that

$$m^y_{k,i} := \tilde{y}_i(k) \bmod N, \quad m^z_{k,i} := \tilde{z}_i(k) \bmod N,$$

where $\tilde{y}_i(k) = q(y_i(k))$ and $\tilde{z}_i(k) = q(z_i(k))$.

Then, given the plain text m^y_k, the quantized value can be obtained using

$$\tilde{y}_{k,i} = \begin{cases} m^y_{k,i}, & \text{if } m^y_{k,i} < N/2, \\ m^y_{k,i} - N, & \text{if } m^y_{k,i} > N/2. \end{cases}$$

In this part, we will show that a certain value bounds the Mean Square Error (MSE) the Kalman filter with the quantization errors. To this end, we first present the following two lemmas to identify the convergence of the Kalman filter without quantization.

Lemma 5.1 *Given the Quantization function* $q(\cdot)$, *the trace of the covariance* $tr(cov(y_k - \tilde{y}_k))$ *is bounded by* $tr(cov(y_k - \tilde{y}_k)) \le n_y/(4\kappa^2)$.

Proof The maximum rounding error can be computed as

$$|\tilde{y}^e_{k,i}| \le \max_{y_{k,i}} |\tilde{y}_{k,i} - y_{k,i}| = \max_{y_{k,i}} |\lceil \kappa \cdot y_{y_{k,i}} \rfloor/\kappa - y_i|$$

$$= \max_{y_i(k)} \underbrace{|\lceil \kappa \cdot y_{k,i} \rfloor - \kappa \cdot y_{k,i}|}_{\le 1/2}/\kappa = 1/(2\kappa).$$

Accordingly, the stochastic variable $\tilde{y}^e_{k,i}$ is in the range $[-1/(2\kappa),1/(2\kappa)]$. Given Popoviciu's inequality on variances, the variance of $\tilde{y}^e_{k,i}$ is bounded by

$$\text{var}(\tilde{y}^e_{k,i}) \le 1/(4\kappa^2). \tag{5.12}$$

Since each element in $\tilde{y}^e_k$ is independent with each other, the trace of the covariance matrix $cov(\tilde{y}^e_k)$ is bounded by

$$\text{tr}(cov(y_k - \tilde{y}_k)) = \text{tr}(cov(\tilde{y}^e_k)) \le n_y/(4\kappa^2). \tag{5.13}$$

Hence we arrive at the result. □

Remark 5.4 Lemma 5.1 shows that we design the upper bound of $\text{tr}(cov(\tilde{y}^e_k))$ by choosing an appropriate scaling factor $\kappa > 0$.

Lemma 5.2 *Define that* $\Sigma_k := \mathbb{E}\|x_k - \hat{x}_k\|^2$. *Suppose that* A_k *and* C_k *are constant matrices, and* (A_k, C_k) *is observable. Then, when* $k \to \infty$, *we have*

$$P_k \to P, \quad H_k \to H, \quad \Sigma_k \to \Sigma, \tag{5.14}$$

where $\Sigma = (I - HC)P$.

The proof of Lemma 5.2 can be found in many literature related to Kalman filter. Lemma 5.2 characterizes the convergence of the Kalman filter without quantization errors. Based on Lemma 5.2, we present the following theorem to identify the upper bound of the MSE with the quantization errors.

Theorem 5.4 *Let $\tilde{x}_{k+1} := A_k\hat{x}_k + H_k(\tilde{y}_k - \tilde{z}_k)$ be the estimated results with quantized inputs $\tilde{y}_k$ and $\tilde{z}_k$. Then, we have the following upper bounded MSE*

$$\lim_{k\to\infty} \mathbb{E}\|(x_k - \tilde{x}_k)\|^2 \le tr(\Sigma) + \frac{n_y tr(HH^T)}{2\kappa^2}.$$

Proof Note that the stochastic variables $(x_k - \hat{x}_k)$ and $(\hat{x}_k - \tilde{x}_k)$ are independent. According to Lemma 5.2, we observe that

$$\mathbb{E}\|(x_k - \tilde{x}_k)\|^2 = \mathbb{E}\|x_k - \hat{x}_k\|^2 + \mathbb{E}\|\hat{x}_k - \tilde{x}_k\|^2$$
$$= tr(\Sigma_k) + \mathbb{E}\|\hat{x}_k - \tilde{x}_k\|^2. \tag{5.15}$$

According to Lemma 5.2, we have

$$\mathbb{E}\|\hat{x}_{k+1} - \tilde{x}_{k+1}\|^2 = \mathbb{E}\|H(\tilde{y}_k^e - \tilde{z}_k^e)\|^2$$
$$\le tr(Hcov(\tilde{y}_k^e)H^T) + tr(Hcov(\tilde{z}_k^e)H^T) \le \frac{n_y tr(HH^T)}{2\kappa^2}. \tag{5.16}$$

Combining (5.15) and (5.16) yields the result. $\qquad\qquad\qquad\qquad\qquad\qquad\qquad\square$

Remark 5.5 Given the quantization errors, Theorem 5.4 implies the MSE is greater than that without quantization, but the MSE is still bounded by a function of the scaling factor κ. To achieve desired performance, we can select a significant scaling factor κ, mitigating the impact caused by the quantization.

5.6 Experimental Results

In this section, we use experiments to illustrate the idea of the proposed mechanism. We study a pollution sensing network in a large-scale thermal power system, since monitoring the concentration of the pollution is critical to public health. We first show how the Paillier cryptosystem encrypts the sensing information. Then, we also present the impact of the quantization errors on the performance of the data assimilation.

5.6.1   The Output of the Encrypted Information

In our previous work [176], we have proposed an environment-aware power generation scheme in smart grids. In the framework, we discretize a particular area into $n_x = 250 \times 200$ points. Each element $x_i(k)$ of the state $x_k \in \mathbb{R}^{n_x}$ stands for the concentration at point i and time k. The matrix A_k characterizes the diffusion of the pollution, and the matrix C_k describes the topology of the sensors. If the number and the position of the sensors are fixed, then C_k is a constant matrix. In our case, we randomly deploy 50 sensors in the area and assume that C_k is constant.

The study period is 24 h, and the sampling time is 30 min. The details parameters of A_k are given in [176]. Suppose that in the Paillier cryptosystem, we choose the large primes as $a = 39989$ and $b = 59999$. In applications, these primes can be much greater than that to make it computationally difficult. Accordingly, we can compute the public key PK and secret key SK defined in Sect. 4.1.

For a given sensor, the time-response of original sensing information $y_i(k)$ and the encrypted information $\bar{y}_i(k)$ are illustrated in Fig. 5.3. In Fig. 5.3a, we observe that the concentration moves in a regular way, while in Fig. 5.3b, the

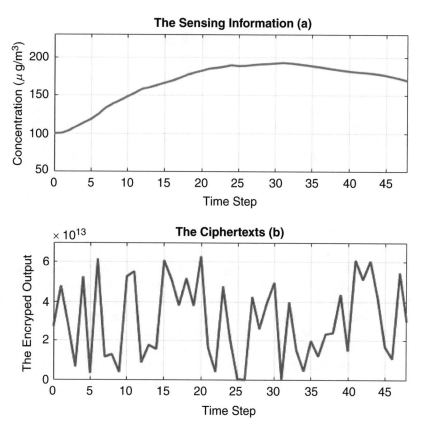

Fig. 5.3 The plaintext and its ciphertexts

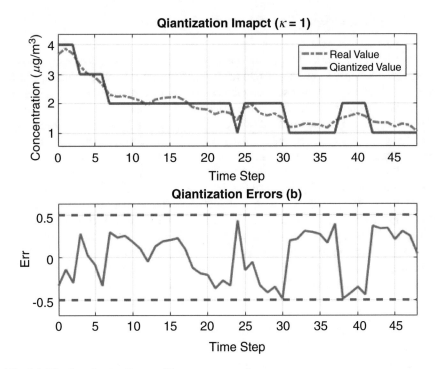

Fig. 5.4 The Quantization Errors with parameter $\kappa = 1$

encrypted output changes randomly. The reason is that the Paillier cryptosystem is not deterministic due to the random number r. Hence, it is also difficult for the adversary to identify the sensing information y_k based on the ciphertexts $\bar{y}_k$. Hence, the proposed mechanism protects the security of the sensing information when each sensor uploads its message to an untrusted cloud.

5.6.2   The Impact of the Quantization Errors

In the second experiment, we verify the quantization error for each sensor is bounded by $1/(2\kappa)$. Figure 5.4 and with $\kappa = 1$ and $\kappa = 5$, respectively. We can see that it is bounded by $1/2$, while in Fig. 5.5, the quantization error is bounded by 0.1. These results coincide with the analytical results in Theorem 5.4. Hence, to reduce the quantization error, we need to choose a large κ and two large primes a and b to guarantee that assumptions in Definition 5.1 is satisfied. Besides, given the bounded quantization error, the mean square error of the Kalman filter is also bounded. This coincides with the results of Theorem 5.4.

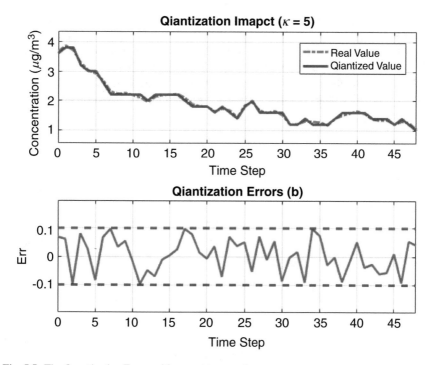

Fig. 5.5 The Quantization Errors with parameter $\kappa = 5$

5.7 Conclusions and Notes

In this work, we have proposed a secure mechanism for large-scale sensor networks to outsource the data assimilation to an untrusted cloud. A combination of standard and customized encryption techniques are used to achieve fully homomorphic property in the outsourcing process. We have analyzed the impact of the quantization errors in the estimation and present an upper bound on the mean-square error. Numerical experiments of a large-scale pollution sensor network in smart grids have corroborated the proposed mechanism.

The techniques proposed in this work apply to many emerging applications, including massive Internet of Things (IoT) networks [46, 47, 47, 150], distributed machine learning [50, 92], and cloud computing [195, 196]. Federated learning [180] is an example of a distributed machine learning technique in which training is conducted across multiple servers or devices that hold their data. The security issues in such a learning mechanism would become important, as investigated in [187–189, 193, 194] for various attack models. Privacy is another concern. For example, in [190], dynamically differential privacy algorithms have been developed to guarantee the privacy of sensitive data that each server has while communicating the training results with other servers. An example of the privacy-preserving distributed machine learning scheme has been applied to connected and autonomous vehicles for machine learning-based intrusion detection systems [192]. Homomorphic encryptions provide a possible technique to ensure the confidentiality and integrity of the computations. Another application is machine learning as a service (MLaaS) [139] that provides machine learning as part of cloud computing services.

Part III
Game-Theoretic Approach for CPS

Chapter 6
Review of Game Theory

6.1 Introduction to Game Theory

Game theory is a mathematical model that investigates interactions among rational and strategic players. These models are essential to our security design for CPSs since we can use them to study the behaviors between defenders and attackers. For example, Backhaus et al. [10] have used game theory models to study the cyber-physical security of the control system with human interactions. Li et al. [94] have used game theory models to develop defense strategies against jamming attacks on remote state estimation of CPSs.

Game theory also plays an essential role in connecting cyber and physical layer problems. In the following subsections, we will introduce three game-theoretic models that are closely related to our applications. In the next chapter, we will show how these models can be applied to the secure and resilient design of CPSs.

6.2 Two-Person Zero-Sum Game Model

In game theory, a zero-sum game is a mathematical representation of a situation, where each player's gain or loss of utility is exactly balanced by that of the other player [114]. Zero-sum games are a specific example of constant sum games (the total utility of the game is always a constant), where the sum of each outcome is always zero. For example, we assume that $U_d \in \mathbb{R}$ is the utility of the defender, and $U_a \in \mathbb{R}$ is the utility of the attacker. If we use a zero-sum game to capture the behaviors between the attacker and defender, then we have

$$U_d + U_a = 0.$$

© The Editor(s) (if applicable) and The Author(s), under exclusive license to
Springer Nature Switzerland AG 2020
Q. Zhu, Z. Xu, *Cross-Layer Design for Secure and Resilient Cyber-Physical Systems*,
Advances in Information Security 81, https://doi.org/10.1007/978-3-030-60251-2_6

Fig. 6.1 The two-player
zero-sum game. The two
interests of the players are
mutually exclusive

Figure 6.1 illustrates the properties of a zero-sum game; i.e., the interests of the players are mutually exclusive. Based on the zero-sum feature, we only need to define one utility function U, i.e., $U = U_d = -U_a$. Then, the defender aims to maximize U, while the attacker aims to minimize U. We will present a formal definition of a zero-sum game in the following subsection.

6.2.1 Formulation of the Zero-Sum Game

In a two-person zero-sum game, we let $\sigma_d \in \Sigma_a$ and $\sigma_a \in \Sigma_a$ be the actions of the defender and attacker, respectively, where Σ_d and Σ_a are action spaces. Given the zero-sum feature, we only need to define shared utility function $U : \Sigma_d \times \Sigma_a \to \mathbb{R}$. The defender aims to maximize the utility function, while the attacker aims to minimize the utility. Hence, we have the following minimax problems:

$$\sigma_a \in \arg\min_{\sigma_a}\max_{\sigma_d}\ U(\sigma_d, \sigma_a),$$

$$\sigma_d \in \arg\max_{\sigma_d}\min_{\sigma_a}\ U(\sigma_d, \sigma_a).$$

The following definition characterizes a Nash equilibrium of the zero-sum game.

Definition 6.1 A pair of strategies (σ_d^*, σ_a^*) constitutes a Nash equilibrium of the zero-sum game if

$$U(\sigma_d, \sigma_a^*) \le U(\sigma_d^*, \sigma_a^*) \le U(\sigma_d^*, \sigma_a), \quad \forall \sigma_d \in \Sigma_d, \sigma_a \in \Sigma_a.$$

We also call (σ_d^*, σ_a^*) as a saddle-point equilibrium.

A zero-sum game is a suitable framework for the security design of CPSs since the attacker can be the minimizer, who aims to decrease the utility of the CPSs, while the defender can be the maximizer, who aims to increase the utility of the system.

6.3   Stackelberg Game Model

A Stackelberg game has two players: one is the leader who takes its action first; the other one is the follower who takes its action after observing the leader's response [169]. To find its optimal strategies, the follower can maximize its utility function based on the action of the leaders. The leader can find its optimal strategy by anticipating the best response of the follower.

We will present a formal definition of the Stackelberg game in the following subsection.

6.3.1   Formulation of the Stackelberg Game

We use L to represent the leader and F to represent the follower. Let $\sigma_L \in \Sigma_L$ and $\sigma_F \in \Sigma_F$ be the strategies of the leader and follower respectively, where Σ_i is the strategy space of player i, for $i \in \{L, F\}$. We define the utility function as $U_i : \Sigma_L \times \Sigma_F \to \mathbb{R}$. Given leader's strategy σ_L, the follower solves the following problem:

$$\mu_F(\sigma_L) \in \arg \max_{\sigma_F \in \Sigma_F} U_F(\sigma_L, \sigma_F),$$

where $\mu_F : \Sigma_L \to \Sigma_F$ is the policy of the follower.

The leader, who can anticipate the policy of the follower, aims to solve the following problem

$$\max_{\sigma_L \in \Sigma_L} U_L(\sigma_L, \mu_F(\sigma_L)).$$

Hence, the equilibrium of the Stackelberg game is given by (σ_L^*, σ_F^*), where

$$\sigma_L^* \in \max_{\sigma_L \in \Sigma_L} U_L(\sigma_L, \mu_F(\sigma_L)), \quad \sigma_F^* = \mu_F(\sigma_L^*).$$

Since the follower can always observe the pure actions taken by the leader, the follower can always choose a pure strategy. By anticipating the follower's pure strategy, the leader can also adopt a pure strategy. According to this property, a finite-action Stackelberg game always admits a pure-strategy equilibrium [13].

6.3.2   Security Design Based on Stackelberg Game

The Stackelberg game plays a significant role in the security design of CPSs. Firstly, we can use the Stackelberg game to capture the relationship between the defender and the attacker. For example, a sophisticated attacker can launch its attack after observing the defense strategy taken by the defender. In this scenario, the defender is the leader, and the attacker is the follower. In another situation, the defender can be the follower, who takes a resilient action after observing the consequence of attacks. In this case, the attacker is the leader who maximizes the damage based on his anticipation of the optimal resilient strategy chosen by the defender.

Besides describing the relationship between a defender and an attacker, we can also use Stackelberg games to capture the interactions between cyber and physical layers. For instance, in a CPS, the cyber layer facilitates data exchange among different physical devices. Before making a decision or implementing a controller, the physical layer needs to wait for data coming from the cyber layer. The nature of this sequential process forms a leader-follower architecture. In this situation, we can also use the Stackelberg game to emphasize the relationship between these two layers.

The Stackelberg game can also contribute to the cross-layer design. In the cross-layer design, we will develop different security mechanisms for different layers. Due to the interdependencies between the cyber and physical layers, these security strategies also depend on each other. Hence, we can use Stackelberg games to find optimal coupling defense strategies. For example, we let the cyber and physical layers be the leader and follower, respectively. Given a specific cyber defense strategy, we design an optimal physical defense policy. After that, we develop an optimal cyber defense policy based on the physical control. In this application, the Stackelberg game achieves the secure cross-layer design holistically.

According to the above discussion, we can see that Stackelberg games facilitate the security design of CPSs. In Chap. 8, we will present a specific application to illustrate the contributions of the Stackelberg game.

6.4   FlipIt Game Model

Recently, the number of targeted attacks has increased significantly in sophistication, undermining the fundamental assumptions on which most cryptographic primitives rely on security. For example, attackers can use an Advanced Persistent Threat (APT) to entirely steal cryptographic keys, violating the assumption that an attacker cannot obtain the key in specific computational time. Hence, the APTs create new challenges, which conventional solutions are not sufficient to address. This motivates us to develop new frameworks to capture the features of APTs.

To protect CPSs from APT attacks, we can update the cyber layer, e.g., change the secret keys of the cryptography, after a successful attack. However, in general, the defender does not know whether the attacker takes over the cyber system of the CPSs; i.e., the defender has incomplete information about the cyber system. One way to solve the issue is to update the cyber layer, e.g., change the key periodically. However, it is difficult to find the optimal frequency of updating the cyber layer. If the updating frequency is high, the costs (e.g., the cyber system needs to pay the

Fig. 6.2 The graphical representation of FlipIt game [167]: The arrows represent the moves of the defender and attacker, and the shaded rectangles present the control of the printing system

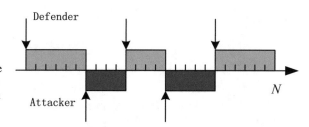

price of the upgrading) of the defender is also high. If the updating frequency is low, then it can cause severe damage to the system when the attacker compromises the cyber layer. Hence, we will introduce a FlipIt game model to capture the situation, where both players aim to control specific resources. In the following subsection, we will introduce the formulation of the FlipIt game.

6.4.1 Formulation of FlipIt Game

We first show a graphical representation of the game in Fig. 6.2. The control of the resource is graphically depicted using the shaded rectangle. A gray rectangle represents a period of defender's control, while a dark red represents a period of attack's control. Players can gain the control of the resources by taking action. We use the arrows to depict the movement of each player in Fig. 6.2.

As illustrated in Fig. 6.2, in a FlipIt game, we have two players, who compete for a shared resource. We use $k = 0, 1, \ldots, N$ to denote the move times, where $N \in \mathbb{Z}_+$ is the length of time. The time-dependent variable $C(k)$ denotes the current player controlling the resource at time k. $C(k)$ is 1 if the player controls the resource, otherwise it is 0. For player i, for $i \in \{a, d\}$, we use G_i to denote its total utility. Then, we have

$$G_i(N) = \sum_{k=0}^{N} C_i(k).$$

Since the total length of the game is N, we have the following relationship

$$G_a(N) + G_d(N) = N.$$

Hence, for each player, we can compute its average gain, given by

$$\gamma_i = G_i(N)/N.$$

It is clear that

$$\gamma_a + \gamma_d = \frac{G_a(N) + G_d(N)}{N} = 1.$$

We let $U_i(N)$ denote the utility of player i up to time N. We assume that player i needs to pay a cost $\alpha_i > 0$ to move. For player i, we let $n_i(N)$ be the total number of the moves. Thus, $U_i(N)$ is given by

$$U_i(N) := G_i(N) - \alpha_i \cdot n_i(N).$$

We let the $\beta_i(N)$ be the average benefit rate up to time N:

$$\beta_i(N) = \frac{U_i(N)}{N} = \frac{G_i(N)}{N} - \alpha_i \cdot \frac{n_i(N)}{N}$$

$$= \gamma_i(N) - \alpha_i \cdot f_i(N). \qquad (6.1)$$

where $f_i(N) := n_i(N)/N$ is the frequency of the moves. In general, we let N go to infinity, i.e., we consider an asymptotic benefit rate, given by

$$\beta_i = \lim_{N \to \infty} \inf \beta_i(N).$$

In the next section, based on the asymptotic benefit rate, we will define the Nash equilibrium of the `FlipIt` Game. Nash equilibrium solutions provide an optimal strategy for the defender to deter an intelligent attacker.

6.4.2   Analysis of the `FlipIt` Game

In the previous subsection, we define the asymptotic benefit rate of each player. The objective of the player is to choose its strategy (e.g., frequency f_i of player i) to maximize its benefit rate. Therefore, we present the following definition to characterize the Nash equilibrium.

Definition 6.2 Let $\mathcal{F}_i$ be the frequency space of player i, for $i \in \{d, a\}$. A Nash equilibrium for the `FlipIt` game is a pair of strategies $(f_d^*, f_a^*) \in \mathcal{F}_d \times \mathcal{F}_a$ such that:

$$\beta_d(f_d^*, f_a^*) \geq \beta_d(f_d, f_a^*), \quad \forall f_d \in \mathcal{F}_d,$$

$$\beta_a(f_d^*, f_a^*) \geq \beta_a(f_d^*, f_a), \quad \forall f_a \in \mathcal{F}_a.$$

Remark 6.1 According to Definition 6.2, we can see that at the equilibrium (f_d^*, f_a^*), each player has no incentive to deviate from its strategy.

Given Definition 6.2, we present the following theorem to characterize a Nash equilibrium based on the unit prices (α_d, α_a) of the move costs.

Theorem 6.1 *The* FlipIt *game, where both players use periodic strategies with random phases, has the following Nash equilibria:*

$$
\begin{cases}
f_d^* = \frac{1}{2\alpha_a}, \ f_a = \frac{\alpha_d}{2\alpha_a^2}, & \text{if } \alpha_d < \alpha_a, \\
f_d^* = f_a = \frac{1}{2\alpha_a}, & \text{if } \alpha_d = \alpha_a, \\
f_d^* = \frac{\alpha_a}{2\alpha_d^2}, \ f_a = \frac{1}{2\alpha_d^2}, & \text{if } \alpha_d > \alpha_a.
\end{cases}
\tag{6.2}
$$

Readers can find the proof of Theorem 6.1 in [167]. In Chap. 7, we will use the FlipIt Game model to design a secure mechanism to protect networked 3D printers from cyber-physical attacks. We will develop an algorithm to compute the optimal strategies of defender based on the results of Theorem 6.1.

6.5 Signaling Game with Evidence

In this section, we will illustrate the basic framework of Signaling Game with Evidence (SGE). Our game model will be slightly different from the one introduced in [174].

In an SGE, we have two players: one is the sender, and the other one is the receiver. The sender has a private identity, denoted by $\theta \in \Theta := \{\theta_0, \theta_1\}$, where θ_0 means the sender is benign, and θ_1 means the sender is malicious. According to its identity, the sender will choose a message $m \in \mathcal{M}$ and send it to the receiver. After observing the message, the receiver can choose an action $a \in \mathcal{A}$. Action $a = 1$ means that the receiver accepts the message, while $a = 0$ means the receiver rejects the message. The sender and receiver have their own utility functions U_i : $\Theta \times \mathcal{M} \times \mathcal{A} \to \mathbb{R}$, for $i \in \{s, r\}$.

Here, given a message $m \in \mathcal{M}$, we assume that both players are aware of the corresponding detection results, i.e., $q = f_q(m)$. Hence, both players' can select the optimal strategies based on detection result q. To this end, we let $\sigma_s(f_q(m)|\theta) \in \Gamma_s$ and $\sigma_r(a|f_q(m)) \in \Gamma_r$ be the mixed strategies of the sender and receiver, respectively. The spaces Γ_s and Γ_r are defined by

$$
\Gamma_s := \left\{ \sigma_s \ \middle| \ \sum_{m \in \mathcal{M}} \sigma_s(f_q(m)|\theta) = 1, \forall m, \ \sigma_s(f_q(m)|\theta) \geq 0 \right\},
$$

$$
\Gamma_r := \left\{ \sigma_r \ \middle| \ \sum_{a \in \mathcal{A}} \sigma_r(a|f_q(m)) = 1, \ \forall a, \ \sigma_r(a|f_q(m)) \geq 0 \right\}.
$$

Note that formation of strategy $\sigma_s(f_q(m)|\theta)$ does not mean that the sender can choose detection results directly. Instead, the sender can only choose message $m \in \mathcal{M}$, which leads to a detection result q based on function $f_q(m)$. For a more complicated case, we can assume that the attacker only obtain an estimated value of q, i.e., $\hat{q}$. We can further build a connection between $\hat{q}$, q, and m. Suppose that attackers has an estimation function $h : \mathcal{Q} \times \mathcal{M} \to \mathcal{Q}$, i.e., $\hat{q} = h(q, m)$.

Given mixed strategies σ_s and σ_r, we define players' expected utility functions as

$$\bar{U}_s(\theta, \sigma_s, a) := \sum_{q \in \mathcal{Q}} \sigma_s(f_q(m)|\theta) U_s(\theta, m, a),$$

$$\bar{U}_r(\theta, m, \sigma_r) := \sum_{a \in \mathcal{A}} \sigma_r(a|f_q(m)) U_r(\theta, m, a).$$

To find the optimal mixed strategies of both players, we define the Perfect Bayesian Nash Equilibrium (PBNE) of the SGE in the following definition.

Definition 6.3 A PBNE of a SGE is a strategy profile (σ_s^*, σ_r^*) and a posterior belief $\pi(\theta)$ such that

$$\sigma_s(f_q(m)|\theta) \in \arg\max_{\sigma_s \in \Gamma_s} \sum_{a \in \mathcal{A}} \sigma_r(a|f_q(m)) \bar{U}_s(\theta, \sigma_s, a),$$

$$\sigma_r(a|f_q(m)) \in \arg\max_{\sigma_r \in \Gamma_r} \sum_{\theta \in \Theta} \pi(\theta) \bar{U}_r(\theta, m, \sigma_r),$$

and the receiver updates the posterior belief using the Bayes' rule, i.e.,

$$\begin{aligned} \pi(\theta) &= f_b(\pi'(\theta), f_q(m)) \\ &:= \frac{\sigma_s(f_q(m)|\theta')\pi'(\theta)}{\sum_{\theta' \in \Theta} \sigma_s(f_q(m)|\theta')\pi'(\theta')}. \end{aligned} \tag{6.3}$$

where $f_b : (0, 1) \times \mathcal{Q} \to (0, 1)$ is the belief-update function, and $\pi'(\theta)$ is a prior belief of θ.

Remark 6.2 Definition 6.3 identifies the optimal mixed strategies of the sender and receiver. Note that at any PBNE, the belief $\pi(\theta)$ should be consistent with the optimal strategies, i.e., at the PBNE, belief $\pi(\theta)$ is independent of time k. Instead, $\pi(\theta)$ should only depend on detection results $q \in \mathcal{Q}$. Besides, we implement Bays's rule to deduce belief-update function (6.3).

In a SGE, there are different types of PBNE. We present three types of PBNE in the following definition.

Definition 6.4 (Types of PBNE) An SGE, defined by Definition 6.3, has three types of PBNE:

1. Pooling PBNE: The senders with different identities use identical strategies. Hence, the receiver cannot distinguish the identities of the sender based on the

available evidence and message; i.e., the receiver will use the same strategies with the different senders.

2. Separated PBNE: Different senders will use different strategies based on their identities, and the receiver can distinguish the senders and use distinct strategies for different senders.

3. Partially-Separated PBNE: different senders will choose different, but not completely opposite strategies, i.e.,

$$\sigma_s(f_q(m)|\theta_0) \neq 1 - \sigma_s(f_q(m)|\theta_1).$$

Remark 6.3 In the separated PBNE, the receiver can obtain the identity of the senders by observing a finite sequence of evidence and messages. However, in the other two PBNE, the receiver may not be able to distinguish the senders' identity.

Note that in applications, the CPS will run the SGE repeatedly, and generate a sequence of detection results $\mathcal{H}_k := \{q_0, q_1, \dots, q_k\}$. At time k, we define the posterior belief as

$$\pi_k(\theta) := \Pr(\theta|\mathcal{H}_{k-1}).$$

Whenever there is a new detection result q_k, we can update the belief using $\pi_{k+1}(\theta) = f_b(\pi_k(\theta), q_k)$, where function f_b is defined by (6.3). Belief $\pi_{k+1}(\theta)$ becomes a prior belief at time $k + 1$.

6.6 Conclusion and Notes

In this chapter, we have briefly introduced several game-theoretic models that will be used in later chapters for cross-layer design. For example, we can use a zero-sum game to describe an adversary model that captures the incentive, the strategies, and the interactions between the attacker and the defender. The Stackelberg game has been presented to describe the hierarchical interactions between cyber and physical layers. The cyber layer can be viewed as the leader, sending commands to the physical layer, acting as a follower. Motivated by APTs, FlipIt game has been presented to capture essential properties of the APT in which an attacker can fully take over CPS resources. We have also presented signaling games with evidence as an example of games of incomplete information. The private identity of the sender is captured by "type," a random variable unknown to the receiver. The concept of perfect Bayesian Nash equilibrium characterizes the outcome of the game and guides the designs of secure CPS. In the following chapters, we will present specific applications in which we use these game models to design the security mechanism.

This chapter has introduced the fundamentals of game-theoretic methods for cybersecurity. The zero-sum games often serve as benchmark models for attacker-and-defender interactions. The applications of such games have been seen in robust design of robust controllers [11, 201], network security [200], adversarial machine learning [189, 193], jamming attacks on wireless systems [209], and worst-case analysis, for example, in infrastructure protection [63, 64] and network design [30, 33]. Zero-sum games can be naturally extended to nonzero-sum ones, especially when there is a need to capture adversaries' additional features, such as the cost of attacks [208] and the cost of learning [203].

Stackelberg games are a special class of sequential games in which players take turns and play sequentially. Sequential games, in general, are useful to capture the multi-round interactions in adversarial environments. For example, in [206], a multi-stage and multi-phase game has been used to describe the adversarial interactions at different levels of the network. The stages and phases of the game correspond to different steps in the cyber kill chain that goes from reconnaissance to target manipulation. In [157, 212], a similar concept has been applied to multi-hop communication networks that are subject to jamming attacks. Wireless devices and nodes in the network search and determine the most secure multi-hop path that escapes from jammers. It has been shown that game-theoretic learning algorithms can significantly improve the security of wireless networks and their resiliency.

Signaling games are a special class of Bayesian games or games of incomplete information that are particularly important to model situations where players do not have complete information on the game which they are involved in. These situations are pervasive in cybersecurity. The attackers often know more about the defenders than what defenders know about the attackers. The information asymmetry naturally creates an attacker's advantage. Cyber deception is an essential mechanism [128, 129] to reverse the information asymmetry and create additional uncertainties to deter the attackers. Readers can refer to [129] for a recent survey of game-theoretic methods for defensive deception and [197] for a summary of Bayesian game frameworks for cyber deception. The Bayesian games can also be extended to dynamic ones to model multi-stage deception. Interested readers can see [65] for more details.

Chapter 7
A Game-Theoretic Approach to Secure Control of 3D Printers

7.1 New Challenges in Networked 3D Printers

In the world of rapid prototyping and small-scale manufacturing, there are two principal methods. The first one is Subtractive Manufacturing (SM), where a mold is poured, or a block of material is reduced through milling and turning. The second one is Additive Manufacturing (AM), also known as 3D printing, which creates parts and prototypes by fusing layers of material together to build the object [4]. Since the object is created layer-by-layer, AM methods can produce more complicated parts than SM methods can. This key advantage of AM attracts attention from numerous agencies and companies, such as NASA, ESA, and SpaceX, enabling a broad range of applications, e.g., the printing of device components, replacement parts, and houses [179].

The main challenges of using 3D-printing technology to produce high quality and precision components remain expansive, e.g., over $20,000 for high-quality plastics [35]. In addition, based on specific applications, manufactured objects should meet a broad range of physical requirements, e.g., resistance to mechanical or thermal stress. Hence, tuning of multiple manufacturing parameters, the pattern of heat source is necessary. Due to the high cost of 3D-printing infrastructure and special knowledge of tuning of parameters, outsourcing the production to third parties specializing in 3D-printing technology becomes necessary [159].

The development of Information and Communication Technologies (ICTs) facilitates the outsourcing of the 3D-printing process, creating a cyber-physical structure for the 3D-printing system. The wide variety of application areas and the dependency on computerization of the 3D-printing process create an attractive target. Two main security threats exist in the 3D-printing system [179]: intellectual property theft and physical damage. Many information technologies have proposed numerous methods to deal with the first threat, but the second threat has not

© The Editor(s) (if applicable) and The Author(s), under exclusive license to
Springer Nature Switzerland AG 2020
Q. Zhu, Z. Xu, *Cross-Layer Design for Secure and Resilient Cyber-Physical Systems*,
Advances in Information Security 81, https://doi.org/10.1007/978-3-030-60251-2_7

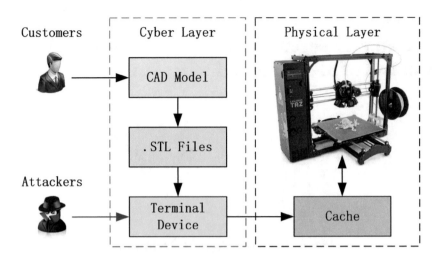

Fig. 7.1 The cyber-physical structure of a 3D-printing system: The adversary can sabotage the system by completely take over the terminal device of the system

been adequately addressed. As 3D-printing systems are critical to military or civil applications (e.g., producing weapons, vehicles, and other hardware), there is a growing desire of attackers and opponents to sabotage the 3D-printing systems. A real example is that a 3D printer explosion occurred in November 2013 at Powderpart Inc. due to mismanagement [158]. Other threats can come from the attacks that target cyber-physical systems; e.g., Stuxnet malware targets Supervisory Control and Data Acquisition (SCADA) system of industrial control systems (ICSs) [109].

In this work, we focus on the cyber-physical attacks for 3D-printing systems. To prevent or deter a cyber-physical attack, we need to understand the vulnerabilities of the 3D-printing system. The cyber-physical nature of the 3D-printing process, shown in Fig. 7.1, provides an opportunity for an attack to infiltrate the physical part of the system. In this process, a user creates a Computer-Aided Design (CAD) model, converts it to a .STL file, and sends the file to the terminal device of the 3D-printing system through networks. The attacker can launch an attack to the 3D-printing system by intruding the terminal device, and modify the .STL file to change the shape of the objects in low resolution, waste the material source, damage the physical components via misoperation [159]. By doing so, the attacker can damage the printing system physically through a cyber intrusion. For instance, the attacker can launch an Advanced Persistent Threat (APT) attack, which can fully steal the secret keys in the terminal device, to access the .STL files stored in the database of the terminal device before sending them to the printer [89].

To protect the system, we aim to design a cross-layer security mechanism for both the cyber and physical layers of the system by applying game-theoretical frameworks. In the physical layer, we develop a zero-sum game to achieve robustness for the printing system. Since the .STL files record important parameters related to the system model [179], the adversary can select malicious parameters to damage the system (e.g., parameters determining the moving speed of the extruder). Hence we use a discrete-time Markov jump system to model these abrupt variations [37]. In the cyber layer, we use a FlipIt game to capture the behaviors between the attacker and the defender of controlling the terminal device. Building on top of that, we use a Stackelberg game to model the interactions between these two layers and define an equilibrium to analyze the effects of the potential attacks. Subsequently, we use simulations to evaluate the performance of the proposed cross-layer mechanism.

7.2 Problem Formulation

In this section, we consider a cyber-physical architecture for the 3D-printing system. We first introduce the dynamics of the 3D-printing system, and use Markov jump systems (MJS) to capture the variability of parameters in the system. Given the system model, we present a cyber attack model and three-game models for different layers.

The proposed framework involves three games as follows. The zero-sum game, denoted by G_Z, takes place between the controller and disturbance; the FlipIt game, denoted by G_F, takes place between the users (or defender), and the attacker; the Stackelberg game, denoted by G_S, takes place between the cyber-layer and physical-layer games.

7.2.1 The Dynamic Model of 3D Printing Systems

Before analyzing the potential threats, we introduce a simplified dynamic model of a 3D-printing system. In general, a 3D-printing system uses stepper rotors to control an extruder to build an object layer by layer. For convenience, we analyze how the rotor moves the extruder in one dimension. The dynamic model of moving the extruder in other dimensions is similar. Figure 7.2 illustrates that a rotor moves an extruder in one dimension via an elastic belt.

Fig. 7.2 The dynamic model of the extruder moving in one dimension: The x_1 is the position of the extruder and the maximum range of x_1 is L

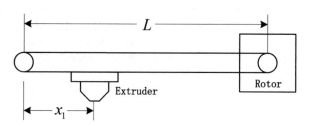

Fig. 7.3 The linearized model of the rotor and extruder: The rotor torque and elastic belt are linearized as two strings with coefficients k_1 and k_2; all damping forces are assumed to be viscous with constants c_1 and c_2

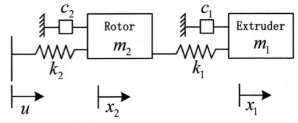

To simplify our problem, we use a linear model to capture the dynamics of the system with following assumptions: (1) rotor torque is modeled as a spring with a constant $k_2 \in \mathbb{R}_+$; (2) the elastic belt is modeled as a linear string with a constant $k_1 \in \mathbb{R}_+$; (3) all damping forces are assumed to be viscous with constants $c_1 \in \mathbb{R}_+$ and $c_2 \in \mathbb{R}_+$. Figure 7.3 shows the simplified linear model.

Using the linearized model shown in Fig. 7.3, we can obtain the following kinetic equations:

$$m_1 \ddot{x}_1 = -k_1(x_1 - x_2) - c_1 \dot{x}_1,$$

$$m_2 \ddot{x}_2 = -k_2(x_2 - u) + k_1(x_1 - x_2) - c_2 \dot{x}_2.$$

Defining $x(t) = [x_1(t), \dot{x}_1(t), x_2(t), \dot{x}_2(t)]^T \in \mathbb{R}^4$, we obtain a continuous-time linear system, given by

$$\dot{x}(t) = A_c x(t) + B_c u(t) + E_c w(t), \tag{7.1}$$

$$x(0) = x_0, \quad x_1(t) \in [0, L],$$

where

$$A_c = \begin{bmatrix} 0 & 1 & 0 & 0 \\ -\frac{k_1}{m_1} & -\frac{c_1}{m_1} & \frac{k_1}{m_1} & 0 \\ 0 & 0 & 0 & 1 \\ -\frac{k_1}{m_2} & 0 & -\frac{k_1+k_2}{m_2} & -\frac{c_2}{m_2} \end{bmatrix}, \quad B_c = \begin{bmatrix} 0 \\ 0 \\ 0 \\ \frac{k_2}{m_2} \end{bmatrix},$$

and $E_c \in \mathbb{R}^{4 \times 1}$; $u(t) \in \mathbb{R}$ is the control input, $w(t) \in \mathbb{R}$ is the disturbance.

The 3D-printing system is naturally a digital control system. Hence we discretize the system with a sampling time T_s. In addition, the linear model (7.1) is obtained by linearizing the original nonlinear model at an equilibrium point. The rotor working at different speeds will alter the system parameters in A_c, B_c and E_c. Hence we use a discrete-time Markov jump system to capture this feature, and it is given by

$$x_{k+1} = A_{\theta_k} x_k + B_{\theta_k} u_k + E_{\theta_k} w_k, \quad x_0 = x_{\text{int}}, \tag{7.2}$$

where $k \in \mathbb{Z}_+$, $x_k = x(kT_s)$, $u_k = u(kT_s)$, and $x_{int} \in \mathbb{R}^4$ is a given initial condition. $\theta_k \in \Theta$ is a Markov process with a given initial condition $\theta_0 \in \Theta$ and the following transition probability:

$$\Pr(\theta_{k+1} = j | \theta_k = i) = \lambda_{ij}(p_a)$$
$$= \Pr(\theta_{k+1} = j | \theta_k = i, \xi = a) p_a$$
$$+ \Pr(\theta_{k+1} = j | \theta_k = i, \xi = d)(1 - p_a),$$

where p_a is the proportion of time that the attacker controls the system. $\xi \in \{d, a\}$ is the index of the cyber player, where d is the defender, and a is the attacker. $\xi = a$ means the case where the attacker takes hold of the device, while $\xi = d$ means the case where the defender has its legitimate control of the device. The conditional probability $\Pr(\theta_{k+1} = j | \theta_k = i, \xi)$ can be obtained by analyzing history information or reinforcement learning. Note that $\Pr(\theta_{k+1} | \theta_k, a)$ is the transition under an assumed attack. This can be known based on attack models or learned through history data. For the case of $\Pr(\theta_{k+1} | \theta_k, d)$, the transition is due to the reliability of the device under proper maintenance. Reliability of the system can be computed through the reliability of its constituent components, which can be found in product specifications or manuals.

In our case, we define set $\Theta := \{0, 1\}$, where $\theta = 0 \in \Theta$ means that the extruder moves in a low-speed mode, and $\theta = 1 \in \Theta$ means that the extruder moves in a high-speed mode. For $\theta_k \in \Theta$, matrices $A_{\theta_k} \in \mathbb{R}^{4 \times 4}$, $B_{\theta_k} \in \mathbb{R}^{4 \times 1}$, and $E_{\theta_k} \in \mathbb{R}^{4 \times 1}$ are derived by discretizing the model (7.1) under different working modes.

7.2.2 Physical Zero-Sum Game Framework

The trajectory tracking problem at the physical layer can be designed through a formulation of an optimal control problem. A summary of optimal control theory has been provided in Appendix B for interested readers. After receiving the .STL file, the printing system converts it to a tool path file, which can be viewed as a reference trajectory (x^r, u^r) for the extruder. Due to the existence of disturbance, we use a zero-sum game framework to develop an optimal robust controller for the printing system.

The robust control problem is formulated by the following min-max problem [11]:

$$\min_{u} \max_{w} J(u, w, x^r, u^r) := \mathbb{E}_\theta \{L(x, u, w, x^r, u^r; \theta)\}, \tag{7.3}$$

with the cost function $L(x, u, w; r)$ given as

$$L(x, u, w, x^r, u^r; \theta) = \sum_{k=0}^{T-1} c(x_k, u_k, w_k, x_k^r, u_k^r; \theta_k) + q_f(x_T; \theta_T),$$

where $T \in \mathbb{Z}_+$ is the horizon length of the problem; $c(x_k, u_k, x_k^r, u_k^r, w_k; \theta_k) :=$ $\|x_k - x_k^r\|_{Q_{\theta_k}}^2 + \|u_k - u_k^r\|^2 - \gamma^2\|w_k\|^2$ is the one-stage cost at time k, $\|\cdot\|$ denotes the $\mathcal{L}_2$-norm, and $Q_{\theta_k} \in \mathbb{R}_+^{4\times4}$ are positive and symmetric for $\theta_k \in \Theta$; $x_k^r \in \mathbb{R}^{4\times1}$, $u_k^r \in \mathbb{R}$ are the reference trajectories for state and control at time k; $\gamma > 0$ is a given threshold; $q_f(x_T; r_T)$ is the terminal cost.

The robust optimal control goal is to solve a zero-sum game to achieve an optimal pair of policy (μ^*, ν^*) for the problem (7.3), where $\mu \in \mathcal{U}_{FB}$ is an admissible feedback loop control strategy, and $\nu \in \mathcal{W}_{FB}$ is the strategy of the disturbance. $\mathcal{U}_{FB}$ and $\mathcal{W}_{FB}$ are the set of the strategies of the controller and disturbance.

7.2.3   A Cyber-Physical Attack Model for 3D-Printing Systems

Cyber-physical nature inevitably brings new security challenges to the 3D-printing system [117]. Many cyber-physical vulnerabilities of the 3D-printing system have been discussed in the literature, e.g. see [159]. Potential threats to 3D-printing systems can come from different parts of the process, as shown in Fig. 7.1, but the most vulnerable component is the .STL file stored in the terminal device. The attacker can intrude on the terminal device to modify the file, resulting in a low-grade or toxic object.

In this work, we consider Advanced Persistent Threats (APTs) type of attacks [167]. In this type of attacks, the attacker can exploit the vulnerabilities of the 3D-printing system and completely take over the terminal device of the printing system (e.g., stealing the password of the terminal device). As the .STL files are usually stored in the database of the terminal device for a while, it is possible for the attacker to access the files and modify them. The .STL files generally determine the tool-path files for the printer, which means modifying a .STL file is equivalent to providing a new trajectory (x^r, u^r) for the extruder. Malicious trajectories or mismanagement may damage the printing system physically, create an unanticipated impact on the defender (or administrator) of the system, and lead to a huge cost. These are the consequences we aim to prevent or mitigate. Besides fabricating the reference (x^r, u^r), the attacker can also alter the moving speed of the extruder by modifying the corresponding parameters in the .STL files [159], resulting in changing the system dynamics (e.g., changing the system matrices A_c, B_c, and E_c). It is the main reason we apply a Markov jump system to capture this feature.

In order to change the .STL files, the attacker needs to spend efforts to break into the terminal device. Here, we call this effort as a move cost, denoted by $\alpha_a \in \mathbb{R}_+$. The defender can mitigate the attacker's effect by updating the passwords or security strategies, leading to another move cost, denoted by $\alpha_d \in \mathbb{R}_+$. Besides changing the passwords, the defender can also improve the security level of the terminal device, bringing a high move cost α_a for the attacker.

7.2.4 The Cyber FlipIt Game Model

At the cyber layer, we apply a FlipIt game framework to model the interactions between the defender, denoted by d, and attacker, denoted by a, under the proposed cyber-attack model. The FlipIt game is a two-player game with a shared resource that both players want to control [167]. In our scenario, we assume that a user wants to produce N number of identical objects. Normally, the user only needs to send the .STL file and the quantity of the objects to the printing system once. However, the attacker can break into the terminal device to modify the .STL file by paying a cost α_a. The defender can recover terminal device by upgrading the system and paying a cost α_d. The attacker also needs to pay a cost for breaking into the terminal device.

For each printing mission, we denote the unit utility of the defender by $U_d \in \mathbb{R}_+$. Let $J_d \in \mathbb{R}_+$ and $J_a \in \mathbb{R}_+$ be the printing cost under the control of defender and attacker. In $\mathbf{G}_F$, the goal of the defender is to maximize its utility U_d^F, given by

$$U_d^F = (U_d - J_d)l_d - \alpha_d h_d.$$

The goal of the attacker is to maximize the printing cost to damage the system. Hence, the utility of attacker is

$$U_a^F = J_a l_a - \alpha_a h_a.$$

l_ξ is the length of time when the player ξ controls the printing system, h_ξ is the number of the moves taken by the player ξ. Let p_ξ be the proportion of time that player ξ controls the printing system, then the average utility rate β_ξ can be obtained by

$$\beta_d := U_d^F/N = (U_d - J_d)p_d - \alpha_d f_d, \tag{7.4}$$

$$\beta_a := U_a^F/N = J_a p_a - \alpha_a f_a, \tag{7.5}$$

where $p_\xi := l_\xi/N$ and $f_\xi := h_\xi/N$, for $\xi \in \{d, a\}$. The defender and attacker aim to seek a pair of strategies (p_a, p_d) to maximize their average utility rates β_d and β_a, respectively.

In this work, the FlipIt game $\mathbf{G}_F$ takes place periodically, and both players' strategies are non-adaptive; i.e., each player does not know who controls the terminal device until it pays the move cost α_ξ [167]. Later, we will define a Nash equilibrium of the FlipIt game, and present a result demonstrating the equilibrium strategies of the attacker and defender.

7.2.5 A Cyber-Physical Stackelberg Game Model

After defining the cyber and physical games, we define a cyber-physical Stackelberg game G_S to compose these two games. The reasons that we use Stackelberg game framework are twofold. Firstly, the cyber-physical interactions are sequential in nature. The 3D-printing system waits for the .SLT file or reference commands before it prints. The defender or attacker can choose their actions by estimating the behaviors of the printer. Therefore, the Stackelberg game framework captures the two-stage dynamic process. Secondly, the players at the cyber and physical layers are often not coordinated; i.e., the network administrator who manages the network assets does not design physical layer controllers, and vice versa. Therefore, the cyber-physical design problem cannot be tackled in a centralized fashion.

In the Stackelberg game, the physical layer is the follower, who chooses a pair of strategies (μ, ν) after observing the actions (p_a, p_d) of the leader; the cyber layer is the leader, who decides (p_a, p_d) by anticipating the optimal respond $J_\xi^*(p_a)$ of the follower [12]. Figure 7.4 illustrates the architecture of the cyber-physical Stackelberg game. The cyber layer plays the leader's role, where the inputs are the strategies p_d and p_a of the user and attacker, respectively, and the outputs are the reference x^r and the attack ratio p_a. The physical layer plays the follower's role, who designs a robust controller based on the x^r and p_a. We present a definition of this cyber-physical Stackelberg game as follow:

Definition 7.1 In this cyber-physical Stackelberg game $\mathbf{G}_S$ defined by (7.3)–(7.5), where the cyber-layer is the leader, and physical layer is the follower, a Stackelberg equilibrium is a strategy pair $\{(p_a^*, p_d^*), (\tilde{\mu}^*, \tilde{\nu}^*)\}$ if it satisfies the following:

$$p_a^* \in \arg\max_{p_a} \beta_a(p_a, p_d^*, \hat{J}_a^*(p_a, p_d^*), \hat{J}_d^*(p_a, p_d^*)),$$

$$p_d^* \in \arg\max_{p_d} \beta_d(p_a^*, p_d, \hat{J}_a^*(p_a^*, p_d), \hat{J}_d^*(p_a^*, p_d)),$$

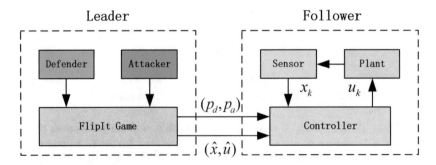

Fig. 7.4 The architecture of the cyber-physical Stackelberg game: the cyber layer is the leader, and the physical layer is the follower

$$\tilde{\mu}^*(p_a, p_d) \in \mathcal{U}_{FB}^*$$

$$:= \{\tilde{\mu}^* \in \mathcal{U}_{FB} | \tilde{J}(\tilde{\mu}^*, \tilde{v}^*) \le \tilde{J}(\tilde{\mu}, \tilde{v}^*), \forall \tilde{\mu}^* \ne \tilde{\mu}\},$$

$$\tilde{v}^*(p_a, p_d) \in \mathcal{V}_{FB}^*$$

$$:= \{\tilde{v}^* \in \mathcal{V}_{FB} | \tilde{J}(\tilde{\mu}^*, \tilde{v}^*) \ge \tilde{J}(\tilde{\mu}^*, \tilde{v}), \forall \tilde{v}^* \ne \tilde{v}\},$$

with

$$\hat{J}^*(p_a, p_d) = g \circ \tilde{J}(\tilde{\mu}^*(p_a, p_d), \tilde{v}^*(p_a, p_d)),$$

where $g : \mathbb{R}_+ \rightarrow \mathbb{R}_+$ is a continuous and monotonically decreasing function that converts the control cost $\tilde{J}$ to the printing cost $\hat{J}$, $\tilde{\mu}^* : [0, 1] \times [0, 1] \rightarrow \mathcal{U}_{FB}$, and $\tilde{v}^* : [0, 1] \times [0, 1] \rightarrow \mathcal{W}_{FB}$.

7.3 Analysis of the Cyber-Physical Games

In the section, we analyze the solutions and the strategies of the players in different games. We also analyze how these three games interact with each other.

7.3.1 Analysis of the Physical Zero-Sum Game Equilibrium

In this subsection, we will solve the zero-sum game problem (7.3) to achieve an optimal control policy $\mu : \mathbb{R}^4 \times \Theta \rightarrow \mathbb{R}$. Define that $e_k := x_k - x_k^r$ and $u_k^e := u_k - u_k^r$, To solve the problem, we apply a dynamic programming approach. Given $\theta_{k+1} = j$ and $\theta_k = i$, we define a value function,

$$V^*(e_k, i, p_a)$$

$$:= \min_{u_k^e} \max_{w_k} \left\{ \tilde{c}(e_k, u_k^e, w_k; \theta_k) + \sum_{j \in \Theta} \lambda_{ij}(p_a) V^*(e_{k+1}, j, p_a) \right\}. \quad (7.6)$$

We assume that $V(e_k, i) = e_k^T P_{i,k} e_k$, where $P_{i,k} \in \mathbb{R}^{4 \times 4}$ is positive for each $i \in \Theta$. Then, we can get

$$V(e_k, i, p_a) = \tilde{c}(e_k, u_k^e, w_k; i) + e_{k+1}^T S_{j,k+1}(p_a) e_{k+1},$$

where $e_{k+1} = A_i e_k + B_i u_k^e + E_i w_k$, $S_{i,k+1}(p_a) := \sum_{j \in \Theta} \lambda_{ij}(p_a) P_{j,k+1}(p_a)$, and $\tilde{c}(e_k, u_k^e, w_k; i) := \|e_k\|_{Q_{\theta_k}}^2 + \|u_k^e\|^2 - \gamma \|w_k\|^2$. Now, we consider the first-order necessary condition:

$$\frac{\partial V(e_k, i, p_a)}{\partial u_k} = 0, \quad \frac{\partial V(e_k, i, p_a)}{\partial w_k} = 0. \tag{7.7}$$

The following inequality,

$$\gamma^2 I - E_i^T S_{i,k+1}(p_a) E_i > 0, \tag{7.8}$$

must be satisfied for all $i \in \Theta$, so the problem (7.6) is strongly concave in w_k for each k. The γ determines the upper bound of the disturbance attenuation in the system (7.2) [11]. A small γ that satisfies (7.8) indicates that the system spends more energy to achieve robustness, leading to a high-quality object. This will incur a high attack ratio p_a because the system is more valuable to intrude.

According to (7.7), we can get

$$u_k^* = K(\theta_k, p_a) e_k + u_k^r,$$
$$w_k^* = L(\theta_k, p_a) e_k,$$

where

$$\begin{aligned}
K(\theta_k, p_a) = \{&I + B_i^T S_{i,k+1}(p_a) B_i - B_i^T S_{i,k+1}(p_a) E_i \\
&\times (E_i^T S_{i,k+1}(p_a) E_i - \gamma^2 I)^{-1} E_i^T S_{i,k+1}(p_a) B_i\}^{-1} \\
&\times \{B_i^T S_{i,k+1}(p_a) E_i (E_i^T S_{i,k+1}(p_a) E_i - \gamma^2 I)^{-1} \\
&\times E_i^T S_{i,k+1}(p_a) A_i - B_i^T S_{i,k+1}(p_a) A_i\},
\end{aligned} \tag{7.9}$$

$$\begin{aligned}
L(\theta_k, p_a) = \{&E_i^T S_{i,k+1}(p_a) E_i - \gamma^2 E_i^T S_{i,k+1}(p_a) B_i \\
&\times (B_i^T S_{i,k+1}(p_a) B_i + I)^{-1} E_i^T S_{i,k+1}(p_a) B_i\}^{-1} \\
&\times \{E_i^T S_{i,k+1}(p_a) B_i (B_i^T S_{i,k+1}(p_a) B_i + I)^{-1} \\
&\times B_i^T S_{i,k+1}(p_a) A_i - E_i^T S_{i,k+1}(p_a) A_i\}.
\end{aligned} \tag{7.10}$$

We update the $P_{i,k}(p_a), \forall i \in \Theta$, using the following coupled Riccati equations,

$$\begin{aligned}
P_{i,k}(p_a) = \; &\phi_i(S_{i,k+1}(p_a), Q_i) \\
:= \; &Q_i + A_i^T S_{i,k+1}(p_a) A_i \\
&- \left[A_i^T S_{i,k+1}(p_a) B_i \quad A_i^T S_{i,k+1}(p_a) E_i \right] \\
&\times \begin{bmatrix} I + B_i^T S_{i,k+1}(p_a) B_i & B_i^T S_{i,k+1}(p_a) E_i \\ E_i^T S_{i,k+1}(p_a) B_i & E_i^T S_{i,k+1}(p_a) E_i - \gamma^2 I \end{bmatrix}
\end{aligned}$$

$$\times \begin{bmatrix} B_i^T S_{i,k+1}(p_a) A_i \\ E_i^T S_{i,k+1}(p_a) A_i \end{bmatrix}, \quad \forall i \in \Theta, \tag{7.11}$$

where $\phi : \mathbb{R}^{4\times4} \times \mathbb{R}^{4\times4} \to \mathbb{R}^{4\times4}$. Note that if E_i is a zero matrix or $\gamma = \infty$, Eq. (7.11) will reduce to the following coupled discrete-time algebraic Riccati equations,

$$P_{i,k}(p_a) = Q_i + A_i^T S_{i,k+1}(p_a) A_i^T - A_i^T S_{i,k+1}(p_a)$$
$$\times B_i (I + B_i^T S_{i,k+1}(p_a) B_i)^{-1} B_i^T S_{i,k+1}(p_a) A_i.$$

The following proposition presents the monotonic properties of the coupled discrete-time algebraic Riccati equations defined by (7.11).

Proposition 7.1 *For any fixed p_a, the $P_{i,k}(p_a)$ is always monotonic in k. In addition, if $T \to \infty$, then, $P_{i,k}(p_a)$ converges to $P_{i,k}^*(p_a)$, where*

$$P_{i,k}^*(p_a) = \phi_i \left(\sum_{j \in \Theta} \lambda_{ij}(p_a) P_{j,k+1}^*, Q_i \right), \quad \forall i \in \Theta.$$

Remark 7.1 The proof of Proposition 7.1 can be found in [3]. According to this proposition, if the attack ratio p_a and the reference (x^r, u^r) are fixed, and T is sufficiently large, then the optimal control cost $J(\mu^*, \nu^*, x^r, u^r)$ is the same.

Then, for a given reference $z^r := [x^r, u^r]^T$ and a p_a, the optimal total control cost is given by

$$\bar{J}(\mu^*, \nu^*; z^r, p_a) := e_0^T P_{\theta_0}^*(p_a) e_0,$$

where $P_{\theta_0}^*(p_a)$ can be recursively computed using (7.11).

We assume that the printing cost is an exponential function of the control cost $\bar{J}$, which is given by

$$J_\xi(p_a) := \delta \exp(-\bar{J}(\mu^*, \nu^*; z^{r_\xi}, p_a)), \tag{7.12}$$

where z^{r_ξ} is the desired trajectory of player $\xi \in \{d, a\}$, and $\delta > 0$ is a coefficient. Based on the printing cost function (7.12), we can see that a small control cost $\bar{J}(\mu^*, \nu^*; z^{r_\xi}, p_a)$ leads to a high printing cost because the smaller control cost results in a better printing performance. The exponential emphasize the margin effect in the printing quality; i.e., the printing cost grows exponentially when one increases the printing performance.

Here, we define a mapping $T_Z : [0, 1] \to \mathbb{R}_+ \times \mathbb{R}_+$. For a fixed attack ratio p_a, we can use the mapping T_F to derive the optimal printing costs (J_d^*, J_a^*) for defender and attacker, respectively, i.e.,

$$\{(J_d^*, J_a^*)\} = T_Z(p_a).$$

7.3.2   Analysis of the Cyber `FlipIt` Game Equilibrium

Given the `FlipIt` game model in Sect. 7.2.4, we define the following Nash equilibrium.

Definition 7.2 A Nash equilibrium of the game $\mathbf{G}_F$ is a strategy profile (p_b^*, p_a^*) such that

$$p_a^* \in \arg\max_{p_a} \beta_a(p_a, p_d^*, J_a, J_d),$$

$$p_d^* \in \arg\max_{p_d} \beta_d(p_a^*, p_d, J_a, J_d).$$

Base on the Definition 7.2, the following theorem characterizes the Nash equilibrium of the `FlipIt` Game.

Theorem 7.1 *For a pair fixed cost* (J_a, J_d)*, the* `FlipIt` *game* $\mathbf{G}_F$*, where both players apply periodic non-adaptive strategies, admits the following equilibria:*

$$p_a^* = \begin{cases} \dfrac{\alpha_d}{2\alpha_a} \cdot \dfrac{J_a}{U_d - J_d}, & \text{if } \dfrac{\alpha_d}{U_d - J_d} \leq \dfrac{\alpha_a}{J_a}, \\[2ex] 1 - \dfrac{\alpha_a}{2\alpha_d} \cdot \dfrac{U_d - J_d}{J_a}, & \text{if } \dfrac{\alpha_d}{U_d - J_d} > \dfrac{\alpha_a}{J_a}, \end{cases} \qquad (7.13)$$

$$p_d^* = 1 - p_a^*.$$

Remark 7.2 Theorem 7.1 presents a closed-form solution of the optimal strategies (p_a^*, p_d^*) for attacker and defender. It is easy to compute the optimal strategies (p_a^*, p_a^*) when the printing costs (J_d^*, J_a^*) are given.

The proof of 7.1 can be found in [167]. According to Theorem 7.1, we define a mapping $T_F : \mathbb{R}_+ \times \mathbb{R}_+ \to [0, 1]$, which maps the printing costs J_d and J_a to the attack ratio p_a. The mapping is given by

$$p_a = T_F(J_d^*, J_a^*).$$

We will employ T_F as part of the definition of an overall equilibrium for $\mathbf{G}_S$.

7.3.3   Analysis of the Cyber-Physical Stackelberg Game Equilibrium

In this subsection, we analyze the solution of Stackelberg game $\mathbf{G}_S$. The following proposition presents an equilibrium of this cyber-physical Stackelberg Game.

Proposition 7.2 (Cyber-Physical Stackelberg Game Equilibrium) *In this cyber-physical Stackelberg game, given the zero-sum game problem (7.3), the printing cost (7.12), and the optimal attacker ratio (7.13), a pair* $\{(p_a^*, p_d^*), (\mu^*, \nu^*)\}$ *is called a Stackelberg equilibrium strategy if*

$$p_a^* = T_F(J_a^*(\mu^*, v^*), J_d^*(\mu^*, v^*)),$$

$$\mu^* = K(\theta, p_a^*) + u_k^r, \ v^* = L(\theta, p_a^*).$$

where $K(\theta, p_a)$ and $L(\theta, p_a)$ are given by (7.9) and (7.10), respectively.

Note that a fixed-point p_a^* satisfies that

$$p_a^* = T_F \circ T_Z(p_a^*). \tag{7.14}$$

Define a mapping $T_S = T_F \circ T_Z : [0, 1] \to [0, 1]$. We give the following theorem to identify the existence of the fixed point p_a^*.

Theorem 7.2 *The mapping T_S is continuous in $[0, 1]$, and T_S has at least a fixed point p_a^* satisfying*

$$p_a^* = T_S(p_a^*).$$

Proof Proving that T_S is continuous in p_a is equivalent to proving the following:

1. $\lambda(p_a)$ is continuous in p_a;
2. $S_{i,k} = \sum_{j \in \Theta} \lambda_{ij} P_{j,k+1}$ is continuous in λ_{ij};
3. $\phi_i(S_{i,k}, Q_i)$ is continuous in $S_{i,k}$;
4. $J_\xi(P_{i,k})$ is continuous in $P_{i,k}$;
5. $T_F(J_a, J_d)$ is continuous in J_a and J_d, respectively.

It is evident that (1), (2), (3) and (4) are satisfied. (5) can also be verified. Therefore, $T_S \in [0, 1]$ is continuous for $p_a \in [0, 1]$. According to Brouwer fixed-point theorem [52], T_S has at a fixed point $p_a^* \in [0, 1]$. □

Corollary 7.1 *Given a p_a^* satisfying $p_a^* = T_S(p_a^*)$, the cyber-physical Stackelberg game $\mathbf{G}_S$ admits an equilibrium $\{(p_a^*, p_d^*), (\mu^*, v^*)\}$, which satisfies*

$$p_d^* = 1 - p_a^*, \quad \mu^* = K(\theta_k, p_a^*)e_k + u_k^r, \quad v^* = L(\theta_k, p_a^*)e_k,$$

where $K(\theta_k, p_a)$ and $L(\theta_k, p_a)$ are defined by (7.9) and (7.10), respectively.

Based on Theorem 7.2, we know that p_a^* always exists. Then, we develop Algorithm 1 to compute the fixed point p_a^*. Algorithm 2 presents an overall cross-layer algorithm for the 3D-printing system.

Remark 7.3 If Algorithm 1 terminates, the solution p_a^* is the fixed point of the mapping T_S. Then, the system can compute the equilibrium solutions using Corollary 7.1.

Remark 7.4 In Algorithm 2, the defender will verify its average utility β_d. When p_a^* is too large, the average utility β_d can be 0. If this happens, the defender should terminate the system and upgrade the security level of the system since it may incur a high risk.

Algorithm 1 Compute the attack ratio p_a^*

1: *Given an initial condition* p_a^0.
2: *Start the following steps iteratively.*
3: *Compute* : $p^{\tau+1} = T_S(p^\tau)$
4: **if** $|p_a^{\tau+1} - p^\tau| < \epsilon$ **then**
5:      **goto** *step* 7.
6: **else goto** *step* 2.
7: *Set* $p_a^* = p_a^{\tau+1}$.

Algorithm 2 Cross-layer algorithm

1: Initializing $U_d, J_d, J_a, \alpha_d, \alpha_a$.
2: **loop**: Cyber layer (*from 1 to N*)
3:      *Determine the reference* (x^r, u^r).
4:      *Update* p_a *using (7.13)*
5:      *Verify the average utility* β_d^F *using (7.4)*
6:      **if** $\beta_d^F = 0$ **then**
7:          *Terminate the system.*
8:      **else**
9:          *Keep the system working.*
10: **loop**: Physical Layer ($k = 1$ to T)
11:      *Accept the reference* (x^r, u^r) *and* p_a.
12:      *Compute optimal policy* μ^* *using (7.9).*
13:      *Compute the printing cost* J_ξ *using (7.12).*
14: **Goto** *step* 2.

7.4   Numerical Results

In this section, we use a 3D-printing system model to illustrate our cross-layer design. The system can be linearized at different equilibria due to the moving speed of the extruder, leading to different linear models. We can obtain the parameters of different models using system identification. For simplicity, we only present two models: low-speed and high-speed models. The parameters of different models are given in Table 7.1 [172].

Based on the parameters given in Table 7.1, we obtain the two linear discrete-time models with a sampling time $T_s = 0.1$ s. The transition probability matrices are given by

$$\Lambda_d = \begin{bmatrix} 0.99 & 0.01 \\ 0.98 & 0.02 \end{bmatrix}, \ \Lambda_a = \begin{bmatrix} 0.2 & 0.8 \\ 0.1 & 0.9 \end{bmatrix},$$

where $\Lambda_\xi(i, j) = \Pr(\theta_{k+1} = j | \theta_k = i, \xi)$, for $\xi \in \{d, a\}$. In these transition matrices, we can see that if the defender (or administrator) controls the printer, then the system has a higher rate working at the low-speed mode. However, if an attacker controls the printer, it has a higher rate working on the high-speed mode, which is

Table 7.1 System parameters of the 3D printing systems

Parameters	X-axis		Y-axis	
	Low speed	High speed	Low speed	High speed
m_1	0.13 kg	0.12 kg	0.13 kg	0.12 kg
m_2	0.45 kg	0.31 kg	1.07 kg	1.05 kg
k_1	110 kN/m	180 kN/m	87 kN/m	101 kN/m
k_2	355 kN/m	410 kN/m	392 kN/m	432 kN/m
c_1	112 Ns/m	130 Ns/m	71.5 Ns/m	86.5 Ns/m
c_2	1100 Ns/m	1350 Ns/m	411 Ns/m	670 Ns/m

harmful to the objects and the system. The weighting matrices Q_0 and Q_1 are given by

$$Q_1 = Q_2 = \text{diag}[100, 1, 10, 1]. \tag{7.15}$$

We define that $\alpha := \alpha_a/\alpha_d$, which is the cost ratio between the attacker and defender. We let the unit utility $U_d = 1$ and the coefficient $\delta = 1$.

In the first experiment, the objects that the printer creates is a cylinder. For convenience, we can focus on a 2D trajectory, e.g., a circle. The reference trajectories at X-axis and Y-axis for the normal user and the attacker are given by

$$\begin{bmatrix} x_1^{r_d} \\ \dot{x}_1^{r_d} \\ y_1^{r_d} \\ \dot{y}_1^{r_d} \end{bmatrix} = \begin{bmatrix} 0.6\cos(0.2\pi t) \\ -0.12\pi\sin(0.2\pi t) \\ 0.6\sin(0.2\pi t) \\ 0.12\pi\cos(0.2\pi t) \end{bmatrix}, \quad \begin{bmatrix} x_1^{r_a} \\ \dot{x}_1^{r_a} \\ y_1^{r_a} \\ \dot{y}_1^{r_a} \end{bmatrix} = \begin{bmatrix} 0.6\cos(0.5\pi t) \\ -0.3\pi\sin(0.5\pi t) \\ 0.6\sin(0.5\pi t) \\ 0.3\pi\cos(0.5\pi t) \end{bmatrix}.$$

It is evident to see that the attacker's trajectory changes more rapidly than the user's does. We set the attack ratio as $p_a = 0.2$. Figures 7.5 and 7.6 illustrate the 2D trajectories drawn by the extruder under normal and malicious trajectories. The results show that the tracking performance is much worse in the Attack case than that in the No Attack case. When working on the high-speed mode, the extruder fails to tracks the trajectory, incurring a high control cost. Figure 7.7 shows the control inputs of the X-axis rotor under different cases. It is obvious that the extruder needs to spend more energy in the attack case since the input amplitude is higher.

In the following experiment, we assume that no security mechanism is applied to the printing system; i.e., we design the controller under the case that $p_a = 0$. Then, we test system performance with a real attack ratio $p_a = 1$. Figure 7.8 shows that when the printing system does not apply the proposed mechanism, the system output performance at X-axis is undesirable. Therefore, without estimating the attack ratio p_a, the system is vulnerable when an attack happens. Figure 7.8 shows that with security mechanism, the system can update p_a^* and track the trajectory better to protect the extruder (Fig. 7.9).

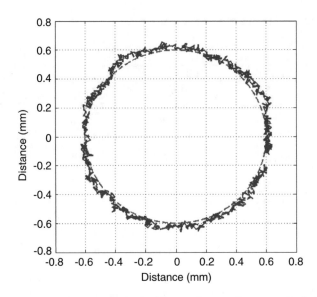

Fig. 7.5 The trajectory of the normal user: In this case, the extruder moves slowly and tracks the trajectory smoothly

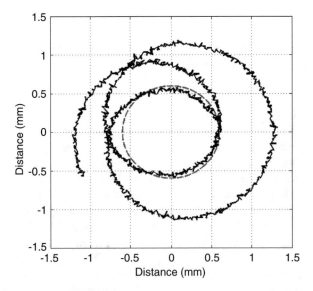

Fig. 7.6 The trajectory of the malicious user: In this case, the extruder moves in a high speed mode, and fails to track the trajectory

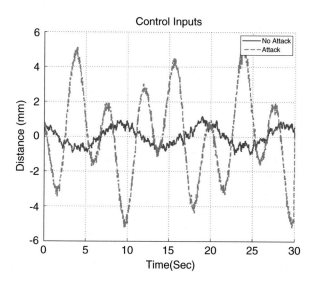

Fig. 7.7 The control input under different cases: The control input of No Attack case is much smaller than that of the Attack case

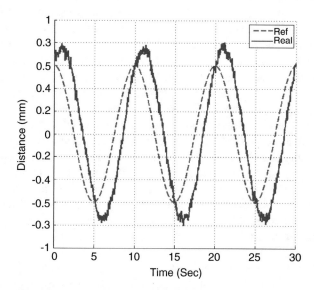

Fig. 7.8 No security mechanism: When an attack happens, if the 3D-printing system remains $p_a = 0$, then the system fails to track the trajectory

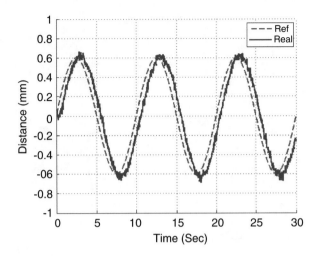

Fig. 7.9 With a security mechanism: When attack happens, with $p_a = 0.6$, the system can track the trajectory

In the next step, we conduct another two case studies. In the first one, we conduct three trials with $\alpha = 1.2, 1.5$, and 2.0. In each trial, we increase γ from γ^* to 10, where γ^* is the smallest γ that satisfies condition (7.8). In the second one, we conduct two trials with $\alpha = 1.3$ and 0.7. In each trial, we run Algorithm 2 with different initial conditions $p_a^0 = 0.2, 0.5$, and 0.8 to see the convergence of the p_a.

Figure 7.10 illustrates the results of the first case study, where the red-dash vertical line indicates the value of γ^*. For each trial, we can see that the smaller the γ is, the higher the attack ratio will be. A small γ indicates that the system achieves a high level of robustness, leading to a high quality of the object. These results show that the attacker has a higher incentive to attack the system when the object has a higher quality. In addition, we can see that the defender can decrease the attack ratio p_a^* by enhancing the cost ratio of α. The cost ratio of α is closely related to the security of the networks. Therefore, the defender can improve the security level of the networks to reduce the attack ratio of p_a. When γ is sufficiently large, the control problem defined by (7.3) reduces to a Linear Quadratic Regulator (LQR) problem. The attack ratio p_a^* converges to a certain point that the printing system uses an LQR controller.

Figure 7.11 illustrates the results of the second case study. In Fig. 7.11, we can see that different initial attack ratios of p_a converge to the same equilibrium p_a^* in different trials. In addition, different cost ratios α lead to different equilibria. When $\alpha = 1.3$, the attack ratio p_a^* is less than 0.5, which means the attacker has a weaker incentive to attack the system. However, when $\alpha = 1.3$, the attack ratio p_a^* is greater than 0.5, which means the attacker has a higher incentive to attack the system.

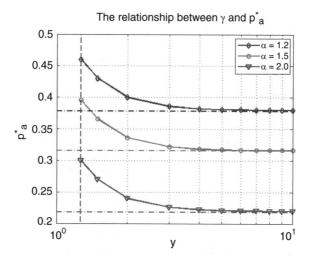

Fig. 7.10 Different values of γ lead to different attack ratios p_a^*: (1) When γ increases, p_a^* decreases because the system is less valuable for the attacker; (2) When α increases, p_a^* decreases because the attack needs to pay a high cost to attack the system

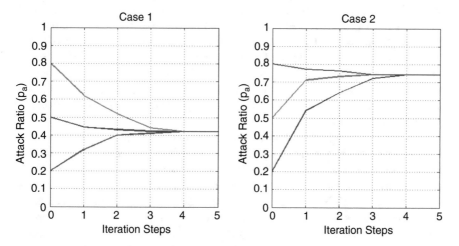

Fig. 7.11 The attack ratios converge using Algorithm 2 under different p_a^0

7.5 Conclusion and Notes

The integration of a 3D-printing system with networked communications constitutes a cyber-physical system, bringing new security challenges to the system. In this work, we have explored the vulnerabilities of the 3D-printing system and designed a cross-layer approach for the system. In the physical layer, we have used a Markov jump system to model the system dynamics and developed a robust control policy to deal with uncertainties. In the cyber-layer, we have used the `FlipIt` game to model the interactions between a user and an attacker. A Stackelberg game is used to integrate the two layers and capture the interactions between these two layers. As a result, we have defined a new equilibrium concept and developed an iterative algorithm to compute the equilibrium. From the conducted experiments, we have observed the impact of the cybersecurity on the physical layer control performances, and an intricate relationship between the security at the cyber layer and the robustness at the physical layer.

This chapter has considered an APT-type threat model where an attacker can launch a stealthy and persistent attack that targets certain assets of the network. We have seen that this attack has lead to failure probabilities unanticipated by the physical layer controller. The mismatch between the anticipated failure rate and the one under adversarial environments has led to degraded control performance. To make the system more resilient, we have designed attack-aware controllers that can use control effort to compensate for the misbehaviors at the cyber layer. From the cyber defense perspective, the security design has been made impact-aware by incorporating the physical consequences into the cyber layer. The impact-aware cyber-defense and the attack-aware controller achieve security and resiliency through a co-design process driven by a Stackelberg game framework. More precisely, the proposed game framework has a games-in-games structure, in which the cyber layer game is nested in the physical layer game. This design philosophy has been generalized in [65, 204] and is applicable to many cyber-physical systems including autonomous vehicles [32] and energy systems [201, 202]. This chapter presents a timely application of 3D-printing systems. It is possible to consider different attack models, e.g., jamming attacks [211] and denial of service [143], and use the co-design methodology for designing secure and resilient mechanisms under a multitude of attacks.

APTs are one important class of attacks on modern CPS. They often rely on a series of sophisticated hacking techniques to gain access to a system and stay in the system for a long period of time to find opportunities to create destructive consequences. The `FlipIt` game considered in this chapter is at an abstract and high level as it mainly captures the consequences of the game and abstracts the attack and defense actions into coarse-resolution "effort" variables. This type of modeling is sufficient for understanding the impact of the APTs on the cyber assets and the ensuing impact on the physical layer. Another way to model APT is to a fine-grained model that captures the attacker's multi-stage kill chain. See [67] and [135, 206] for example. The choice of the APT models would depend on the applications and the design questions to address.

Chapter 8
A Game Framework to Secure Control of CBTC Systems

8.1 Introduction to CBTC Systems

Railway transportation systems play a significant role in critical infrastructures since it moves people, fuel, food, raw materials, and products routinely [38]. The rapid growth of the city and population entails the massive capacity of railway transportation. To meet the growing demand, operators focus on maximizing the train line capacity and keeping the system safe by evolving and adapting the train signaling system.

The development of information and communication technologies (ICTs) facilitates the advent of a modern railway signaling and control system, Communication-Based Train Control (CBTC) system. A CBTC system is an automatic train control system integrated with wireless communications [147]. By exchanging the information among different trains via wireless communications, CBTC systems can increase the capacity of a train line by reducing the time interval (or headway) between trains travailing along the line [171]. One application of CBTC systems is the European Railway Traffic Management System, where each train reports its position and speed to the train movement authority periodically [113].

Figure 8.1 illustrates the architecture of a CBTC system, where each train continuously senses and sends its status (position and velocity) via wireless communications to a wayside access point deployed along the railway. The train can also acquire its preceding train's information [210]. Given its preceding train's information, the train can track its front train's trajectory while keeping a safe distance. By designing an appropriate safe distance, the operators can maximize the capacity of a train line as well as maintain its safety. For security concerns, the safe distance between two trains should be greater than the emergency brake length of the trains.

© The Editor(s) (if applicable) and The Author(s), under exclusive license to
Springer Nature Switzerland AG 2020
Q. Zhu, Z. Xu, *Cross-Layer Design for Secure and Resilient Cyber-Physical Systems*,
Advances in Information Security 81, https://doi.org/10.1007/978-3-030-60251-2_8

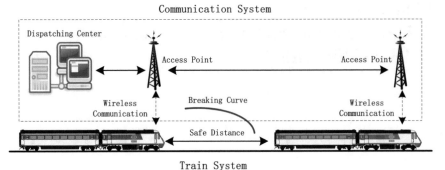

Fig. 8.1 The CBTC system: two trains communicate with each other via wireless links to the time interval between trains travailing along the line

Despite many advantages of the CBTC system, new challenges arise as the integration with wireless communications may expose the system to cyber threats. CBTC systems adopt advanced mobile technologies based on IP-medium access to improve transmitting efficiency, but it is easy for cyber attackers to interfere with the communications [97]. Hence, the potential cyber-security issues in these new train control systems need to be adequately reviewed. To address security issues, we need to take into account the cyber-physical nature of the CBTC system. A CBTC system is a typical cyber-physical system (CPS), where the cyber layer has the wireless communication, and the physical layer contains the train control system. The cyber-physical property provides opportunities for cyber attackers to damage the train system by infiltrating the cyber layer of CBTC systems. Since the control performance of a train significantly relies on the accuracy of its preceding train's information, such as position and speed, any cyber attack that substantially increases the uncertainty of the critical information may cause severe consequences, such as emergency brakes or collisions between two trains.

One possible threat that can damage the physical part of a CBTC system is the jamming attack, where an adversary installs a jamming device on the train to send a high interference to jam the wireless communications [90]. Figure 8.2 illustrates a jamming attack on the CBTC system. By doing so, the jammer can decrease the signal-to-interference-plus-noise ratio (SINR) of the communication system on the train, resulting in a low packet arrival rate (PAR). The small PAR will lead to ambiguity of the front train's information, and hence cause instability or severe accidents of the train system.

This work focuses on jamming attacks on a CBTC system. To mitigate the jamming impact, we consider a multi-channel framework for the communications of the CBTC system. Given the framework, the transmitter of each train can randomly select a channel to send the message to avoid the jamming signal. However, an intelligent attacker can learn the strategy of the transmitter and choose a possible channel to jam. To describe the performance of the communication, we define a

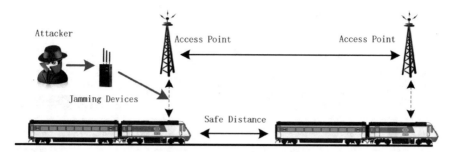

Fig. 8.2 Jamming Attack on CBTC System: an attacker uses a jamming device to interfere with the transmitting signals of the CBTC system

cyber state and apply a stochastic game (SG) to model the dynamics of the cyber state and the behaviors of the transmitter and jammer. The transition kernel of the SG depends on the strategies of the transmitter and jammer. We use dynamic programming to find the equilibrium of the SG. The main contribution of this work is summarized as follows:

1. We investigate the security issues in the CBTC system and analyze the impact of the jamming attack on the physical parts of the system. To capture the uncertainty of the estimation, we identify the minimum safe distance under a given security level.
2. To build a connection between the communication and the estimation, we define an MDP model to describe the dynamics of the cyber state. Based on the MDP, we develop a zero-sum game to model the interactions of the transmitter and jammer and define the equilibrium of the game.
3. We use a dynamic programming approach to achieve the Nash equilibrium. We investigate a two-channel case to obtain analytical results that interconnect the physical parameters of the train and cyber policies of the players. These results emphasize the cyber-physical nature of the CBTC system.
4. The experimental results coincide with our analysis and show that secure design enhances the reliability of the CBTC system.

8.2 Problem Formulation

In this section, we will formulate the physical problem of the CBTC system, define the communication and attack models, and propose the idea of the defense strategy. Figure 8.3 illustrates the interdependencies among different parts of the mechanism and outlines the steps of our design. In Fig. 8.2, the solid blue line shows the forward interdependencies: the defender uses a security mechanism to meet the safety requirements; the requirements can guarantee the safety of the system under

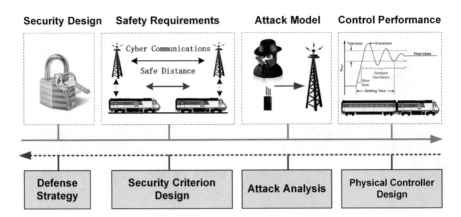

Fig. 8.3 The design steps of the proposed mechanism. The solid blue line shows the forward interdependencies: the defender uses a security mechanism to meet the safety requirements; the requirements can guarantee the safety of the system under the jamming attack and admit a desirable control performance. The dash red arrow shows the backward design: we design a controller for the train, proposed a jamming attack model, define a security criterion based on the safety requirements, and develop a security mechanism to protect the system from the jamming attack

the jamming attack; despite the meeting the safety requirements, the proposed mechanism should also ensure a desirable control performance of the train.

In Fig. 8.3, the dash red arrow outlines the steps of our design. We first introduce a train model and design a controller for the train. Then, we proposed a communication model and a jamming attack model. In the following sections, given the train and attack models, we define a security criterion to meet the safety requirements of the systems. Based on the models and security requirements, we formulate a stochastic game (SG) to capture the interactions between and use the equilibrium of the SG to find the optimal defense strategies.

8.2.1   The Physical Model of a Train System

According to the design steps in Fig. 8.3, we introduce the train model and define a physical control problem. The dynamics of a train system can be modeled as a second-order system. The discrete-time model of a train is given as [171]:

$$z_{k+1} = z_k + \Delta T v_k + \frac{1}{2} \frac{\Delta T^2}{M_{\text{train}}} u_k - \frac{1}{2} \frac{\Delta T^2}{M_{\text{train}}} w_k,$$

$$v_{k+1} = v_k + \frac{\Delta T}{M_{\text{train}}} u_k - \frac{\Delta T}{M_{\text{train}}} w_k,$$

where $\Delta T \in \mathbb{R}_{++}$ is the sampling time; $z_k \in \mathbb{R}$ and $v_k \in \mathbb{R}$ are the position and velocity of the train at time k, respectively; $u_k \in \mathcal{U} \subset \mathbb{R}$ is the control input; $w_k \in \mathbb{R}_+$ is the resistance; $M_{\text{train}} \in \mathbb{R}_{++}$ is the mass of a train. When $u_k > 0$, u_k is the traction effort; when $u_k < 0$, u_k is the brake resistance. In addition, the resistance w_k includes a constant part, e.g., the friction, and a dynamic part proportional to the velocity v_k. Hence, we define w_k as

$$w_k = w_f + c_v v_k,$$

where $w_f \in \mathbb{R}_{++}$ is the friction, and $c_v \in \mathbb{R}_{++}$ a resistance coefficient. Then, we rewrite the discrete-time model into the following form:

$$z_{k+1} = z_k + \left(\Delta T - \frac{1}{2} \frac{\Delta T^2 c_v}{M_{\text{train}}} \right) v_k + \frac{1}{2} \frac{\Delta T^2}{M_{\text{train}}} u_k - \frac{1}{2} \frac{\Delta T^2}{M_{\text{train}}} w_f,$$

$$v_{k+1} = \left(1 - \frac{\Delta T c_v}{M_{\text{train}}} \right) v_k + \frac{\Delta T}{M_{\text{train}}} u_k - \frac{\Delta T}{M_{\text{train}}} w_f,$$

For convenience, we define a discrete-time state variable $x_k = [z_k, v_k]^T \in \mathcal{X} \subset \mathbb{R}^2$ to obtain a linear state-space model, given by

$$x_{k+1} = A x_k + B u_k + E w_f, \quad x_0 \in \mathbb{R}^2, \tag{8.1}$$

where

$$A = \begin{bmatrix} 1 & \Delta T - \frac{1}{2} \frac{\Delta T^2 c_v}{M_{\text{train}}} \\ 0 & 1 - \frac{\Delta T c_v}{M_{\text{train}}} \end{bmatrix}, \quad B^T = \left[\frac{1}{2} \frac{\Delta T^2}{M_{\text{train}}}, \frac{\Delta T}{M_{\text{train}}} \right]^T,$$

$$E^T = \left[-\frac{1}{2} \frac{\Delta T^2}{M_{\text{train}}}, -\frac{\Delta T}{M_{\text{train}}} \right], \tag{8.2}$$

and $x_0 = [z_0, v_0]^T \in \mathcal{X}$ is a given initial condition, and $\mathcal{X} \in \mathbb{R}^2$ is the feasible set of state x_k.

Given the model (8.1), we first consider the control problem with perfect communications. The control objective of a train is to follow its front train's trajectory while keeping a safe distance $L_s > 0$. Let $x_k^f \in \mathbb{R}^2$ be the state vector of the front train. Then, we can define a nominal system

$$x_{k+1}^r = A x_k^r + B u_k^r + E w_f,$$

where $x_k^r := x_k^f - [L_s, 0]^T$ is the reference for the train at time k.

According to the nominal system, we define $e_k^r := x_k - x_k^r$ and $u_k^e := u_k - u_k^r$ as the tracking errors of the state and control, respectively. Hence, the dynamics of the tracking errors can be found using

$$e_{k+1}^r = x_{k+1} - x_{k+1}^r = Ae_k^r + Bu_k^e. \tag{8.3}$$

The train aims to find the optimal control policy $\mu^e : \mathcal{X} \to \mathcal{U}$ to minimize the following cost function:

$$\min_{\mu^e \in \mathcal{U}} J_p(e_0^r, \mu^e) := \sum_{k=0}^{\infty} \left(\|e_k^r\|_Q^2 + \|\mu^e(e_k^r)\|_R^2 \right), \tag{8.4}$$

subject to (8.3), where $\| \cdot \|^2$ is the Euclidean norm; $Q \in \mathbb{R}^{2 \times 2}$ and $P_T \in \mathbb{R}^{2 \times 2}$ are positive symmetric matrices; $R \in \mathbb{R}_{++}$ is a tuning parameter.

We use dynamic programming approach to solve the problem (8.4) and obtain the optimal control u_k^*, given by

$$u_k^* := \mu^e(e_k^r) + u_k^r = Fe_k^r + u_k^r, \tag{8.5}$$

where $F \in \mathbb{R}^2$ is a constant vector. The details of the vector F can be founded in linear control materials [84].

In the case of perfect communication, the train can always receive the accurate information x_k^f at time k. Hence, it is easy to obtain the closed-form solution (8.5). However, communication is not perfect, and we need to construct connections between communications and control problems. In the next subsection, we will introduce communication and attack models. It will be clear that the reliability of wireless communications can affect the control performance of the physical layer.

8.2.2   Communication Model and Attack Model

The multichannel network has shown great promise in network design. One application of multichannel communications to improve communication performance is the E-SAT 300A system, a multichannel satellite communication system for aircrafts [1]. Due to the advantages, we consider a multichannel communication model for the CBTC system.

We apply an MIMO-enabled WLAN with M channels to a CBTC system, but the purpose is to enhance the resiliency of the system to a jamming attack. At each time k, the transmitter can randomly pick a channel to avoid the malicious interference sent by the jammer. In the MIMO model, we assume that each channel has independent additive white gaussian noise (AWGN). Given the Signal-to-Noise Ratio (SNR), we approximate the Packet Arrival Rate (PAR) β_i of the i-th channel, for $i = 1, \ldots, M$, as follows [93]:

$$\beta_i \approx 1 - c_{i,p} \cdot \exp(-c_{i,d} \cdot \Gamma_i), \ i = 1, \ldots, M, \tag{8.6}$$

where $c_{i,p}$ and $c_{i,d}$ are two constant parameters of the ith fading model. Γ_i is the SNR of the ith channel, given by

$$\Gamma_i\Big(p_u(i), 0\Big) = p_u(i)/\sigma_c^2,$$

where $p_u(i)$ is the transmitting power of the ith channel, and σ_c^2 is the variance of the AWGN to the channel.

However, if a jammer chooses the i-th channel to jam with a jamming power $p_a(i)$, then the SNR turns into signal-to-interference-plus-noise (SINR), given by

$$\Gamma_i\Big(p_u(i), p_a(i)\Big) = \frac{p_u(i)}{p_a(i) + \sigma_c^2}.$$

Hence, the corresponding bit error rate and the PAR become

$$\beta_i \approx 1 - c_{i,p} \cdot \exp\Big(-c_{i,d} \cdot \Gamma_i(p_u(i), p_a(i))\Big). \tag{8.7}$$

Figure 8.4 illustrates the architecture of the multi-channel model and jamming attack, where the transmitter has M number of channels to select to send the message at each time k, and the attacker can choose one of them to jam. If the attacker selects the right channel to jam, then PAR will be (8.7). The PAR (8.7) will be significantly lower than that without jamming.

Fig. 8.4 The multi-channel network: the transmitter (resp. the attacker) chooses a channel to transmit (resp. jam)

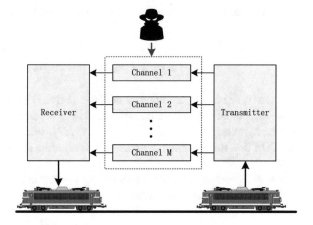

8.3   Estimation Approach and Security Criterion

Due to the unreliability of wireless communication, the train needs to estimate the state x_k^f of the front train when it loses a packet. When the message is received, then we can use it; otherwise, we need to use the system model to estimate the information. In this section, we first present an estimation method. Then, given the estimation, we propose a security criterion for the CBTC system based on the physical security analysis.

8.3.1   Physical Estimation Problem

Note that sampling time ΔT is sufficiently small, and N is not sufficiently large. Hence, we assume that control policy of the front train is stationary, i.e., $u_k^f = F_c x_k^f$, for $F_c \in \mathbb{R}^2$ is a constant row vector, which can be designed by solving the problem (8.4). Accordingly, the estimated value $\hat{x}_k^f$ can be obtained using

$$\hat{x}_k^f = \begin{cases} x_k^f, & \text{if transmission succeeds,} \\ A\hat{x}_{k-1}^f + Bu_{k-1}, & \text{if transmission fails.} \end{cases} \tag{8.8}$$

We do not consider the case that N is sufficiently large, since unexpected events might happen to the front train, altering its control policy.

Given the estimator (8.8), we can obtain the estimation error e_k^f, given by

$$e_k^f = \begin{cases} 0, & \text{if transmission succeeds,} \\ Ae_{k-1}^f + w_{k-1}, & \text{if transmission fails.} \end{cases}$$

Clearly, the larger the number of Consecutive Dropping Packets (CDPs) is, the more inaccurate the estimation will be. Hence, we define θ_k as the number of CDPs at time k, taking from a set $\Theta := \{0, 1, \ldots, N\}$, where N is the highest number of the CDP that the system can endure. We can view θ_k as a cyber state of the cyber layer.

Given θ_k and the physical model (8.1), we define a function $\Omega : \Theta \to \mathbb{R}^{2 \times 2}$ to compute the covariance of estimation error e^f, i.e.,

$$\Omega(\theta) := \mathbb{E}\left[e^f (e^f)^T \middle| \theta\right] = \sum_{l=0}^{\theta-1} A^l \Sigma_w (A^T)^l. \tag{8.9}$$

where $\Sigma_w \in \mathbb{R}^{2 \times 2}$ is the covariance matrix of the Gaussian noise w_k. We observe that the trace of $\Omega(\cdot)$ is increasing in θ, i.e., a large θ implies a poor estimation.

Hence, in Sect. 8.5, we will use the trace of $\Omega(\cdot)$ as a part of the cost function, which the defender (resp. the attacker) aims to minimize (resp. maximize).

8.3.2 Security Criterion for CBTC System

In the previous part, we define a safe distance L_s between two trains, but for security concerns, the safe distance should be longer than the emergency braking distance, denoted by L_b. Then, the margin distance, defined by $L_m := L_s - L_b$, is critical to the security issues. Here, the security issues in the CBTC system are twofold. On the one hand, we ask how long the margin distance L_m should be such that the train system can tolerate up to N number of CDPs. On the other hand, given a L_m, N should be sufficiently large such that the train can keep at least L_m from its proceeding train.

Since the noise to the estimation is stochastic, it is challenging to achieve a perfect security, and the cost of perfect security is expensive. Hence, we define a security level in a probability sense. The following definition characterizes the security of the CBTC system under a given security level.

Definition 8.1 The CBTC system is ε-level secure if

$$\Pr(\hat{z}^f - z_k^f \geq L_m) \leq \varepsilon,$$

where $\varepsilon \in (0, 1)$ is a sufficiently small number.

Remark 8.1 The ε-level security indicates that the probability of difference between the real position z_k^f and the estimated position of the front train greater than L_m is bounded by a negligible scalar $\varepsilon > 0$. Definition 8.1 illustrates that the probability that the train has to take an emergency brake is sufficiently small.

The following theorem indicates the relationship between the margin distance L_m and the largest CDP number N.

Theorem 8.1 *Given L_ϵ and a security level ε such that*

$$\Pr(\hat{z}_k^f - z_k^f \geq L_m) \leq \varepsilon, \tag{8.10}$$

then the maximum packet-dropping number N should be bounded by $N \leq \lfloor \Delta \rfloor$, where $\lfloor \cdot \rfloor$ is the floor function, and Δ is defined by

$$\Delta := N \leq \frac{1}{2} \log_\delta \left(\frac{\varepsilon L_m^2 (\delta^2 - 1)}{\sigma_w^2} + 1 \right), \tag{8.11}$$

and δ is the greatest eigenvalue of the matrix A_c^f.

Proof According to Chebyshev' inequality, we can have

$$\Pr(\hat{z}_k^f - z_k^f \geq L_m) \leq \Pr(|\hat{z}_k^f - z_k^f| \geq L_m) \leq \sigma_z^2/L_m^2, \tag{8.12}$$

where σ_z^2 is the variance of the stochastic variable $\hat{z}^f$. Hence, a sufficient condition to guarantee (8.10) is that

$$\sigma_z^2/L_m^2 \leq \varepsilon \implies \sigma_z^2 \leq L_m^2 \varepsilon. \tag{8.13}$$

According to the matrix norm, the variance σ_z^2 is bounded by

$$\sigma_z^2 \leq \sum_{l=0}^{N} \delta^2 \sigma_w^2 = \sigma_w^2 \frac{1 - (\delta^2)^N}{1 - \delta^2}, \tag{8.14}$$

where δ is the spectral radius of matrix A_c^f. In the case that $\delta > 1$, we observe that

$$\sigma_z^2 \leq \sigma_w^2 \frac{\delta^{2N} - 1}{\delta^2 - 1} \leq L_m^2 \varepsilon$$

$$\Leftrightarrow N \leq \frac{1}{2} \log_\delta \left(\frac{\varepsilon L_m^2 (\delta^2 - 1)}{\sigma_w^2} + 1 \right).$$

Similarly, we can show the same results with $\delta < 1$. This completes the proof. □

Remark 8.2 Theorem 8.1 presents the requirement of safe distance L_s under a given security level. The reason we define the security in a probability sense is that the perfect security can introduce a high cost, which is not worthy since attackers may not exist. In Sect. 8.5, we will show that ε be sufficiently trivial.

The following corollary shows the lower bound of the margin distance L_m when N is given.

Corollary 8.1 *Given N and the security level ε, the margin distance L_m should be bounded by*

$$L_m \geq \sqrt{\sigma_w^2 (\delta^{2N} - 1)/\varepsilon(\delta^2 - 1)}.$$

Corollary 8.2 *If the train operator aims to achieve perfect security, i.e., $\varepsilon \to 0$, then the margin distance L_m is*

$$\lim_{\varepsilon \to 0} L_m = \infty.$$

Remark 8.3 The perfect security is achievable if only one train moves between two stations. However, the capacity of the system stays at its minimum level.

After setting up the physical layers, we will propose a cyber stochastic game problem in the following section.

8.4 The Stochastic Game-Theoretic Framework

In this section, we develop a zero-sum stochastic game, denoted as $\mathbf{G}_Z$, to capture the strategic interactions between the transmitter and jammer. Given the game model, we define a Nash equilibrium (NE), characterizing the optimal strategies of the transmitter and an intelligent attacker. Then, we use a Markov chain to model the dynamics of the cyber state. We apply dynamic programming to find the equilibrium and present analytical results of the problem.

8.4.1 Cyber Zero-Sum Game

In the jamming attack, the attacker (denoted as a) aims to increase the state θ_k to deteriorate the estimation, while the user or transmitter (denoted as u) aims to decrease the state θ_k to achieve the desired estimation. Accordingly, we formulate a stochastic zero-sum game $\mathbf{G}_Z$, where both players share a common cost function J_c. We assume that both players can obtain the parameters of the cost function J_c since the players can learn these parameters through past observations [90].

The game $\mathbf{G}_Z$ is played as follows: at each time k, the transmitter chooses a channel $\eta_d = i$, for $i \in \mathcal{M} := \{0, 1, \ldots, M\}$, to send the message, where $i = 0$ means not to send; the attacker chooses a channel $\eta_a = j$, for $j \in \mathcal{M} := \{0, 1, \ldots, M\}$, to jam, where $j = 0$ means not to jam. $\eta_d, \eta_a \in \mathcal{M}$ are the pure strategies of the transmitter and jammer, respectively. We let $\phi_d(i)$ (or $\phi_a(i)$) be the probability that the transmitter (or attacker) chooses the i-th channel to transmit (or jam). At each time k, we consider the mixed strategies $\phi_d = [\phi_d(1), \ldots, \phi_d(M)]^T \in \Phi_d$ and $\phi_a = [\phi_a(i), \ldots, \phi_a(M)]^T \in \Phi_a$ of the transmitter and attacker, respectively, where Φ_d and Φ_a are the set of the admissible strategies defined by

$$\Phi_d := \left\{ \phi_d \in [0, 1]^{M+1} \,\middle|\, \sum_{i=0}^{M} \phi_d(i) = 1 \right\},$$

$$\Phi_a := \left\{ \phi_a \in [0, 1]^{M+1} \,\middle|\, \sum_{j=0}^{M} \phi_a(j) = 1 \right\}.$$

According to the PAR of the communication model in Sect. 8.2.2, we can define a packet arrival rate (PAR) matrix W, given by

$$W = \begin{bmatrix} 0 & 0 & \cdots & 0 \\ \beta_{10} & \beta_{11} & \cdots & \beta_{1M} \\ \vdots & \vdots & \ddots & \vdots \\ \beta_{M0} & \beta_{M1} & \cdots & \beta_{MM} \end{bmatrix}_{(M+1)\times(M+1)} \quad , \tag{8.15}$$

where β_{ij} is the PAR when the transmitter chooses the i-th channel while the attacker chooses the j-th. β_{ii} is the PAR when the attacker jams the correct channel. Due to the nonnegative jamming power p_a, it is clear that for any channel $i \in \mathcal{M}$, we have

$$\beta_{ii} < \beta_{ij}, \forall j \neq i, j \in \mathcal{M}, \tag{8.16}$$

where β_{ii} is the PAR under a jamming attack, and β_{ij} is the normal PAR since the attacker jams the wrong channel.

Combining with the mixed strategies of the transmitter and attack, the expected packet arrival rate (PAR) is given by

$$\gamma(\phi_d, \phi_a) := \phi_d^T W \phi_a, \tag{8.17}$$

where $\gamma(\phi_d, \phi_a)$ is the expected PAR under the mixed strategies ϕ_d and ϕ_a.

Given the strategies ϕ_d and ϕ_a, we can see the transition of the cyber state from θ_k to θ_{k+1} is independent with the history strategies. Hence, we can use a Markov process to capture the transition of the cyber state θ_k. Figure 8.5 shows a Markov model can capture the dynamics of the cyber state θ_k, and the transition probability depends on the strategies of the players. Note that we have a state N. When θ_k goes to the state N, the train takes an emergency brake and the state stays at N, as illustrated in Fig. 8.5. The transition probability of the Markov chain is given by

$$\Pr\left[\theta_{k+1} = j | \theta_k = i, \phi_d, \phi_a\right] = \lambda_{ij}(\phi_d, \phi_a) = \begin{cases} \gamma, & \text{if } j = 0; \\ 1 - \gamma, & \text{if } j = i + 1; \\ 1, & \text{if } i = j = N + 1; \\ 0, & \text{otherwise}; \end{cases}$$

Fig. 8.5 The Markov model of the cyber layer: the state of the Markov model is the number of CDP, and the transition probability depends on the strategies of the players

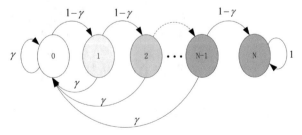

where γ is the short of $\gamma(\phi_d, \phi_a)$. According to Fig. 8.5, the probability that the state θ goes to state N is $P_e = (1 - \gamma)^N$, where P_e is the probability that the train takes an emergency brake.

Let $\{\ell_k\}_{k=0}$ denote the sequence of random cost function, with ℓ_k being the cost for the period $[k, k + 1)$. It should be clear that once an initial state θ and strategies ϕ_d, ϕ_a are specified, then so is the probability distribution of ℓ_k for every $k = 0, 1, \ldots$. Here, we consider a discounted cyber cost function, that is related to the estimation performance under the strategies ϕ_d and ϕ_a. The cyber cost function is defined by

$$J_c(\theta, \phi_d, \phi_a) := \sum_{k=0}^{\infty} \rho^k \mathbb{E}_{(\theta, \phi_d, \phi_a)}\left[\ell_k(\theta_k, \phi_d, \phi_a)\middle|\theta_0 = \theta\right], \qquad (8.18)$$

where $\rho \in (0, 1)$ is a discounting factor.

Let $\pi_\xi : \Theta \to \phi_\xi$ denote a stationary strategy profile of player ξ, for $\xi \in \{d, a\}$. The following definition characterizes a saddle-point equilibrium in the stochastic zero-sum game.

Definition 8.2 A stationary strategy profile $(\pi_d^*, \pi_a^*) \in \Phi_d \times \Phi_a$ is a saddle-point equilibrium of the zero-sum game with a utility function (8.18) if $\forall \theta \in \Theta$, it satisfies

$$J_c(\theta, \pi_d^*, \pi_a) \le J_c(\theta, \pi_d^*, \pi_a^*) \le J_c(\theta, \pi_d, \pi_a^*). \qquad (8.19)$$

Since this game has finite players and actions, the stochastic game admits a saddle-point equilibrium, defined by Basar and Olsder [13]. Figure 8.6 illustrates how the proposed stochastic game captures the cyber-physical structure of the CBTC system. In the next subsection, we will define cost function ℓ_k and find the Nash equilibrium of the game.

8.4.2   Analyzing the Equilibrium of the Game

Note that transmitting and jamming cost energy. Hence, we need to consider the expected power of the transmitter and jammer. Given mixed strategies ϕ_d and ϕ_a, we can obtain the expected power costs $E_d(\phi_d)$ and $E_a(\phi_a)$, defined by

$$E_d(\phi_d) := [0, p_{d,1}, \ldots, p_{d,M}]\phi_d, \quad E_a(\phi_a) := [0, p_{a,1}, \ldots, p_{a,M}]\phi_a,$$

where $p_{d,i}$ and $p_{a,i}$ are the power injected by the transmitter and attacker on the i-channel, respectively.

In this subsection, we will define the cost function $\ell : \Theta \times \Phi_d \times \Phi_a \to \mathbb{R}$ that takes following forms:

$$\ell(\theta_k, \phi_d, \phi_a) := h(\theta_k) + \nu(\phi_d, \phi_a),$$

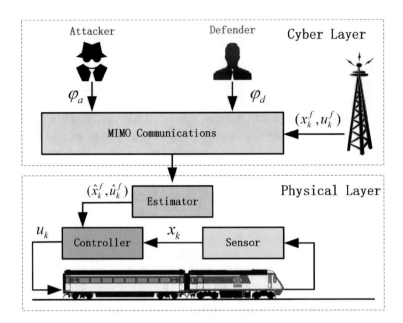

Fig. 8.6 The cross-layer architecture of the proposed mechanism: at the physical player, we design an estimator and controller to achieve a desirable control performance for the train; at the cyber layer, we develop a stochastic game to find the optimal defense strategy ϕ_d while anticipating the optimal attack strategy ϕ_a of an intelligent attacker

with $h(\theta_k) := tr(\Omega(\theta_k))$ and the energy function

$$v(\phi_d, \phi_a) := q_u E_u(\phi_d)\gamma(\phi_d, \phi_a) - q_a E_a(\phi_a)\bar{\gamma}(\phi_d, \phi_a),$$

where $\bar{\gamma} := 1\gamma$; $q_u, q_a \in \mathbb{R}_+$ are the proportions of the energy term in the cost function.

To find equilibrium (8.18), we use value iteration. Define the value function $V(\theta)$ such that

$$V^*(\theta_k) = \min_{\phi_d} \max_{\phi_a} \mathbb{E}\left[\sum_{\tau=t} \rho^{t-k}\ell(\theta_t, \phi_d, \phi_a)\right].$$

We can use the value iteration to obtain the optimal value function V^*, where the iteration is presented as follows:

$$V^*_{\tau+1}(i) = \min_{\phi_d} \max_{\phi_a}\left\{\ell(i, \phi_d, \phi_a) + \rho \sum_{j\in\Theta} \lambda_{ij}(\phi_d, \phi_a)V_\tau(j)\right\}.$$

Based on the contraction mapping theorem, we can show that $V_\tau(\cdot)$ converges to an optimal one V^* as $\tau \to \infty$ [16]. Then, we can derive the optimal policy by solving the following matrix game:

$$\min_{\phi_u} \max_{\phi_a} \quad \phi_u^T G(\theta)\phi_a, \tag{8.20}$$

where the matrix $G : \Theta \to R^{M \times M}$ is defined by

$$G(i) := \left[\ell(i, \eta_d, \eta_a) + \rho \sum_{j \in \Theta} \lambda_{ij}(\eta_d, \eta_a) V^*(j) \right]_{\eta_d, \eta_a = 0}^{M, M}, \quad \forall i \in \Theta. \tag{8.21}$$

In the following, we will show the analytical results of the our problem. The first one is the monotonicity of the function $V_\tau(\theta)$ in θ. To show the monotonicity, we need the following lemma.

Lemma 8.1 *Given two matrices $D, \bar{D} \in \mathbb{R}^{n \times m}$, whose elements satisfy that $d_{ij} \leq \bar{d}_{ij}$, for $i = 1, \ldots, n$, and $j = 1, \ldots, m$, then the value of the matrix game D is no greater than that of the matrix game $\bar{D}$, i.e.,*

$$\min_{y_1} \max_{y_2} y_1^T D y_2 \leq \min_{y_1} \max_{y_2} y_1^T \bar{D} y_2,$$

where $y_1 \in [0, 1]^n$ and $y_2 \in [0, 1]^m$ are vectors of the mixed strategies.

The proof of Lemma 8.1 can be founded in [13]. Then, we present one of our main results characterized by the following theorem.

Theorem 8.2 *For all $\theta \in \Theta$, let $V_0(\theta) = 0$ be the initial value function. At each iteration step τ, the value function $V_\tau(\theta)$ is monotonically increasing in $\theta \in \Theta$.*

Proof Based on the definition of $h(\theta)$, we can show that $h(\theta)$ is increasing in θ. In the value iterations, we start from $V_0(\theta) = 0$.

$$V_1(\theta) = \min_{\phi_d} \max_{\phi_a} \left\{ h(\theta) + v(\phi_d, \phi_a) \right\} = h(\theta) + \min_{\phi_d} \max_{\phi_a} v(\phi_d, \phi_a).$$

Then, we can observe that

$$V_1(\theta + 1) - V_1(\theta) = h(\theta + 1) - h(\theta) > 0.$$

Assuming that $V_\tau(\theta)$ is increasing in $\theta \in \Theta$, we need to prove that $V_{\tau+1}(\theta)$ is also increasing in θ. We observe that

$$V_{\tau+1}(\theta - 1) = h(\theta - 1) + \rho V_\tau(\theta) + \min_{\phi_d} \max_{\phi_a} \phi_u^T G_\tau(\theta)\phi_a,$$

$$V_{\tau+1}(\theta) = h(\theta) + \rho V_\tau(\theta + 1) + \min_{\phi_d} \max_{\phi_a} \phi_u^T G_\tau(\theta + 1)\phi_a,$$

where

$$G_\tau(\theta) := \left[v(\eta_d, \eta_a) + \rho \left(\gamma(\eta_d, \eta_a) V_\tau(0) + \bar{\gamma}(\eta_d, \eta_a) V_\tau(\theta) \right) \right]_{\eta_d, \eta_a = 0}^{M}.$$

Since $V_\tau(\theta)$ is increasing in θ, we achieve that $g_{ij}(\theta + 1) > g_{ij}(\theta)$. According to Lemma 8.1, we observe that

$$\min_{\phi_d} \max_{\phi_a} \phi_d^T G_\tau(\theta + 1) \phi_a > \min_{\phi_d} \max_{\phi_a} \phi_d^T G_\tau(\theta) \phi_a.$$

Since $h(\theta)$ and $V_\tau(\theta)$ are both strictly increasing in θ, we note that

$$V_{\tau+1}(\theta + 1) > V_{\tau+1}(\theta), \text{ for } \theta = 0, 1, \ldots, N - 1.$$

This completes the proof.   □

Remark 8.4 The result of Theorem 8.2 is significant to the problem since it provides incentives for the transmitter to move back to the state 0 for a good estimation performance. A good evaluation leads to a desirable control performance. Hence, Theorem 8.2 emphasizes the interdependencies between the cyber and physical layers.

We present the above analytical results based on general cases. In the next subsection, we investigate a particular case to obtain more insightful results.

8.4.3   Special Case Study: Two-Channel Game

In this subsection, we study a two-channel case. We will propose a closed-form solution of the mixed strategies for both players. We explore the inter-dependencies among the physical parameters of the train, cyber policies of the defender and attacker, and the transmitting power and jamming power. These results provide a connection between the cyber and physical layers, indicating that the CBTC has a typical cyber-physical nature.

To show the existence of the closed-form solution, we need the following assumption.

Assumption 8.1 *For any channel $i, j \in \mathcal{M}$, energy function $v(\cdot)$ satisfies that*

$$v(i, j) - v(i, i) < \rho(\beta_{ij} - \beta_{ii})(V(1) - V(0)). \tag{8.22}$$

Remark 8.5 Assumption 8.1 ensures that the difference of the energy costs should be bounded. The main reason is that if (8.22), then we say channel i dominates channel j. We can show that $\phi_{d,j} = 0$. Hence, we can remove channel j from set $\mathcal{M}$ since the defender will not use channel j.

Given Assumption 8.1, the following theorem shows that no pure strategy exists in our stochastic game.

Theorem 8.3 *Based on Assumption 8.1, there is no saddle point (i.e., no pure strategy exists) in zero-sum game $G(\theta)$, defined by (8.21), for any $\theta \in \Theta$.*

Proof Suppose that $G(\theta)$ has a pair of pure strategies $\{\eta_d = j, \eta_a = j\}$; i.e., defender uses channel j to transmit the data, and attacker jams channel i. Since the defender has no incentive to move, we have

$$G_{ii}(\theta) \leq G_{ij}(\theta)$$

$$\Leftrightarrow v_{ii} + \rho\beta_{ii}V(0) + \rho(1 - \beta_{ii})V(\theta + 1) \leq v_{ij} + \rho\beta_{ij}V(0) + (1 - \beta_{ij})V(\theta + 1)$$

$$\Leftrightarrow v_{ii} + \rho\beta_{ii}(V(0) - V(\theta + 1)) \leq v_{ij} + \rho\beta_{ij}(V(0) - V(\theta + 1))$$

$$\Leftrightarrow \rho(\beta_{ij} - \beta_{ii})(V(\theta + 1) - V(0)) \leq v_{ij} - v_{ii}.$$

The last inequality violates 8.1 since $V(\theta+1) - V(0) \geq V(1) - V(0)$. Hence, game $G(\theta)$ has no pure strategy.   □

Due to the non-existence of the pure strategy, we can use the following lemma to compute the closed-form strategies of a two-channel case.

Lemma 8.2 (Owen [119]) *Let $D \in \mathbb{R}^{2\times 2}$ be a 2×2 matrix game. Then if D does not have a saddle point, its unique optimal strategy optimal strategies are given by*

$$\pi_d(\theta) = \frac{[1, 1]D^*}{[1, 1]D^*[1, 1]^T}, \quad \pi_a(\theta) = \frac{D^*[1, 1]^T}{[1, 1]D^*[1, 1]^T}, \tag{8.23}$$

where $D^* \in \mathbb{R}^{2\times 2}$ is the adjoint matrix of D.

Based on Lemma 8.2, we can present how the mixed strategies vary with different cyber states θ. The following theorem shows that in a two-channel case, the optimal mixed strategies can be monotone in θ if the PAR satisfies a given condition.

Theorem 8.4 *Consider a two-channel case; i.e., only channels i and j are available. Then, the optimal mixed strategies $\pi_{d,i}(\theta)$ and $\pi_{a,i}(\theta)$ decrease in θ if*

$$\beta_{ij} - \beta_{ii} > \beta_{ji} - \beta_{jj}. \tag{8.24}$$

Proof Given Theorem 8.3 and Lemma 8.2, we can obtain the closed-form solution of $\pi_{d,i}(\theta)$, given by

$$\pi_{d,i}(\theta) = \frac{J_c(\theta, j, j) - J_c(\theta, j, i)}{J_c(\theta, i, i) - J_c(\theta, i, j) + J_c(\theta, j, j) - J_c(\theta, j, i)} \tag{8.25}$$

where $i, j \in \mathcal{M}$ are the indexes of the channels. Since $\gamma(i, j) = \beta_{ij}$, we can have

$$J_c(\theta, i, j) = h(\theta) + v_{ij} + \rho\left(\gamma(i, j)V^*(0) + \bar{\gamma}(i, j)V^*(\theta + 1)\right)$$

$$= h(\theta) + v_{ij} + \rho V(\theta + 1) + \rho\beta_{ij}\left(V(0) - V(\theta + 1)\right).$$

Hence, we can have

$$J_c(\theta, i, i) - J_c(\theta, i, j) = v_{ii} - v_{ij} + \rho\Delta V^*(\theta)(\beta_{ij} - \beta_{ii}), \qquad (8.26)$$

$$J_c(\theta, j, j) - J_c(\theta, j, i) = v_{jj} - v_{ji} + \rho\Delta V^*(\theta)(\beta_{ji} - \beta_{jj}), \qquad (8.27)$$

where $\Delta V^*(\theta) := V^*(\theta + 1) - V^*(0)$ is increasing in θ due to Theorem 8.2. If $\beta_{ij} - \beta_{ii} > \beta_{ji} - \beta_{jj}$, it is clear that $\pi_{d,i}(\theta)$ is increasing in θ since $v_{ii} - v_{ij}$ and $v_{jj} - v_{ji}$ are constants.

Similarly, we can also show that $\pi_{a,i}^*(\theta)$ is decreasing in θ if (8.24) holds. □

Remark 8.6 When the cyber state θ increases, the transmitter aims to enhance the PAR to send the state back to 0 due to the increasing gap $\Delta V^*(\theta)$. In Theorem 8.4, the inequality (8.24) indicates that the channel i is better if it has a greater difference between the jamming PAR β_{ii} and the normal PAR β_{ij} than that of channel j.

Theorem 8.5 *In a two-channel case, if the attacker increases its power $p_{a,i}$ to power $p'_{a,i}$ for channel i, i.e., $p_{a,i} < p'_{a,i}$, then, the optimal mixed strategy pairs $\{\phi_{d,i}(\theta), \phi_{a,i}(\theta)\}$ and $\{\phi'_{d,i}(\theta), \phi'_{a,i}(\theta)\}$ satisfy that*

$$\phi_{d,i}(\theta) \geq \phi'_{d,i}(\theta), \quad \phi_{a,i}(\theta) \geq \phi'_{a,i}(\theta), \quad \forall\theta \in \Theta.$$

Proof According to the definition of PAR β, defined by (8.7), we can show that

$$\frac{\partial\beta(p_{d,i}, p_{a,i})}{\partial p_{a,i}} < 0, \quad \text{and} \quad \frac{\partial\beta(p_{d,i}, 0)}{\partial p_{a,i}} = 0.$$

Hence, the difference $\beta_{ij} - \beta_{ii}$ is increasing in $p_{a,i}$, while $\beta_{ji} - \beta_{jj}$ remains the same. Hence, according to (8.25)–(8.27), we can deduce that

$$\phi_{d,i}(\theta) \geq \phi'_{d,i}(\theta).$$

Similarly, we can also show that

$$\phi_{a,i}(\theta) \geq \phi'_{a,i}(\theta).$$

This completes the proof. □

Given Theorem 8.5, we can quickly propose the following corollary.

Corollary 8.3 *In a two-channel case, if the defender increases its power $p_d(i)$ to power $p_d'(i)$ for channel i, i.e., $p_d(i) < p_d'(i)$, then, the optimal mixed strategy pairs $\{\phi_{d,i}(\theta), \phi_{a,i}(\theta)\}$ and $\{\phi_{d,i}'(\theta), \phi_{a,i}'(\theta)\}$ satisfy that*

$$\phi_{d,i}(\theta) \le \phi_{d,i}'(\theta), \quad \phi_{a,i}(\theta) \le \phi_{a,i}'(\theta), \quad \forall \theta \in \Theta.$$

Proof The proof of Corollary 8.3 is similar to the proof of Theorem 8.5.   □

Remark 8.7 Theorem 8.5 and Corollary 8.3 build a connection between the transmitting (jamming) power and the cyber policies. We can see that the growing transmitting power of channel i increases the probability of choosing i for both players, while the increasing power of jamming decreases the likelihood.

8.4.4 Inter-Dependency Between Physical and Cyber Layers

In this subsection, we aim to study the relationships between the physical parameters and cyber policies, showing a strong interdependence. We recall that system matrix A, defined by (8.2), has the following structure:

$$A = \begin{bmatrix} 1 & a_1 \\ 0 & a_2 \end{bmatrix}, \quad \text{with } a_1 := \Delta T - \frac{1}{2}\frac{\Delta T^2 c_v}{M_{\text{train}}}, \quad a_2 := 1 - \frac{\Delta T c_v}{M_{\text{train}}}. \quad (8.28)$$

Since the sampling time ΔT is sufficiently small, we assume that a_1 is a constant for different trains, i.e., $a_1 \approx \Delta T$. However, parameter a_2 will vary for different trains since mass M_{train} can be different. Another factor can affect a_2 is the coefficient c_v, which depends on the wind speed and the shape of the train [171].

Then, the following theorem characterizes the relationship between the physical parameters a_2, defined by (8.28), and cyber policies (π_d, π_a).

Theorem 8.6 *In a two-channel zero-sum game $G(\theta)$, cyber polices $\pi_{d,i}(\theta)$ and $\pi_{a,i}(\theta)$ are increasing in physical parameter a_2, defined by (8.28), for any $\theta \in \Theta$ if*

$$\beta_{ij} - \beta_{ii} > \beta_{ji} - \beta_{jj}.$$

Proof Note that

$$A^l = \begin{bmatrix} 1 & s \\ 0 & a_2^l \end{bmatrix}, \quad \text{with } s := \sum_{t=0}^{l-1} a_1 a_2^t.$$

Hence, we can have

$$\Sigma_w (A^T)^l A^l = \begin{bmatrix} \sigma_p^2 & 0 \\ 0 & \sigma_v^2 \end{bmatrix} \begin{bmatrix} 1 & 0 \\ s & a_2^l \end{bmatrix} \begin{bmatrix} 1 & s \\ 0 & a_2^l \end{bmatrix} = \begin{bmatrix} \sigma_p^2 & \sigma_p^2 s \\ \sigma_v^2 s & \sigma_v^2 a_2^{2l} \end{bmatrix}.$$

According to (8.9), we arrive at

$$h(\theta) = tr\left\{\sum_{l=0}^{\theta} \Sigma_w (A^T)^l A^l\right\} = \sum_{l=0}^{\theta} tr\left\{\Sigma_w (A^T)^l A^l\right\} = \sum_{l=0}^{\theta} (\sigma_p^2 + \sigma_v^2 a_2^{2l}).$$

We can conclude that the value of $h(\theta)$ increases when a_2 increases. Similar to the proof of Theorem 8.2, we can show that $V(\theta)$ is increasing in parameter a_2. Combining (8.25)–(8.27), we can deduce that $\pi_{d,i}(\theta)$ is increasing in parameter a_2 for every $\theta \in \Theta$.

Similarly, we can also show that $\pi_{a,i}(\theta)$ is increasing in parameter a_2 for every $\theta \in \Theta$.   □

Remark 8.8 Theorem 8.6 shows a strong inter-dependency between the physical parameters and cyber policies. A train with a greater parameter a_2 means the system is more unstable, i.e., a larger a_2 leads to a larger eigenvalue of matrix A. Theorem 8.6 tells that both players will choose a channel with a larger difference $\Delta\beta_i = \beta_{ij} - \beta_{ii}$ if the front train is more unstable.

Based on the above analytical results, we also show a close connection between the cyber and physical layers of the CBTC system. In the next section, we will use numerical experiments to show how the physical and cyber layers interact.

8.5 Experimental Results

In this section, we evaluate the performance of our proposed mechanism for both cyber and physical layers. The simulation has two parts: we first show the equilibrium of the cyber zero-sum game and the enhancement of the reliability of the communication under our defensive strategies; then, we show the jamming impact on the control performance of the train. Then, we analyze the improvement of physical performance under the proposed protection.

8.5.1 The Results of Cyber Layer

In the first part, we analyze the equilibrium results of the cyber zero-sum game. We consider a two-channel case, where the energy levels of the transmitter and jammer are given by

$$E_u = [0, \ 0.7, \ 1.5], \quad E_a = [0, \ 0.2, \ 0.5].$$

where channel 1 has a lower transmitting power (or jamming power) than channel 2. If there is no attack, the user prefers channel 2 since it has a higher transmitting power, but under the jamming attack, the user will have tradeoffs when choosing these channels.

For the wireless fast-fading model, we consider the following parameters $c_{1,p} = c_{2,p} = 0.8$, $c_{1,d} = 1.2$, and $c_{2,d} = 1.6$. The variance of the transmitting noise is $\sigma_c^2 = 0.25$. The energy coefficients are $q_u = q_a = 1$. According to the setting, the matrix W defined by (8.15) is given by

$$W = \begin{pmatrix} 0 & 0 & 0 \\ 0.81 & 0.50 & 0.81 \\ 0.94 & 0.94 & 0.67 \end{pmatrix}.$$

We consider that $L_m = 20m$, $L_s = 100m$, and the security level $\varepsilon = 10^{-6}$, i.e., $\Pr(\hat{z}_k^f - z_k^f \geq L_m) \leq 10^{-6}$, which is sufficiently small. Given Theorem 8.1, we can compute the upper bound (8.11) of N, which is 10.63. Then, we take $N = 10$. Then, setting $\rho = 0.9$, we run the value iteration for 50 times. Figure 8.7 shows the convergence of the value iteration and the monotonicity of the value function in state θ. This result coincides with Theorem 8.3.

Figure 8.8 shows the results of the optimal mixed strategies of transmitter at different state. We can see that the mixed strategies on channel 2 is increasing in θ. This result coincides with Theorem 8.4 since $\beta_{21} - \beta_{22} < \beta_{12} - \beta_{11}$.

We can also see that without the protection, the maximum PAR under the jamming attack is 0.67. Then, the probability that the train takes an emergency brake is $P_e = (1 - 0.67)^{10} = 1.5 \times 10^{-5}$. However, with the proposed mechanism, the expected PAR under the jamming attack is between 0.77 and 0.79, and the worst-case P_e is $P_e = (1 - 0.77)^{10} = 4.1 \times 10^{-7}$. We can see that without defense strategies, the probability P_e is around 404 times greater than that with defense strategies. In the next subsection, we will run simulations on the physical performance under the jamming attack.

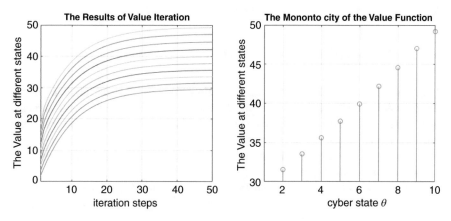

Fig. 8.7 The results of the value iteration, and the monotonicity of the value function

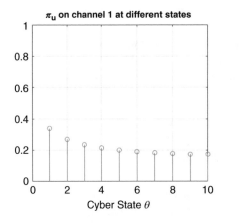
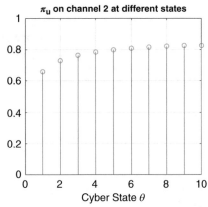

Fig. 8.8 The mixed strategy of the transmitter

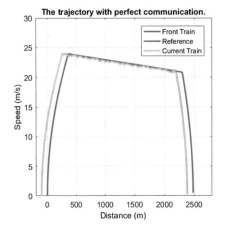
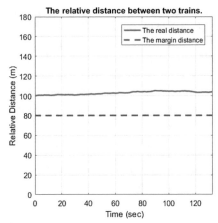

Fig. 8.9 The tracking performance of the train with perfect communication

8.5.2   The Results of Physical Layer

In the second part of the simulation, we evaluate the physical performance under the jamming attack. Figure 8.9 shows the trajectory of the train without attacks, where the y-axis is the speed and x-axis is the position. Figure 8.10 illustrates the outcome with a jamming attack. Due to the jamming impact, the current train moves tremblingly, and it is unsafe since the relative distance is smaller than the emergency braking distance L_b between $k = 61$ and $k = 120$.

Then, we analyze the case with the proposed secure design. In Fig. 8.11, we observe that with the secure design, the relative distance can be guaranteed to be greater than L_b to avoid an emergency brake.

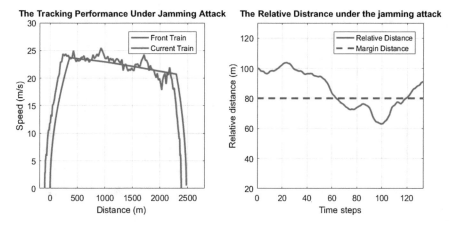

Fig. 8.10 The tracking performance of the train under the jamming attack

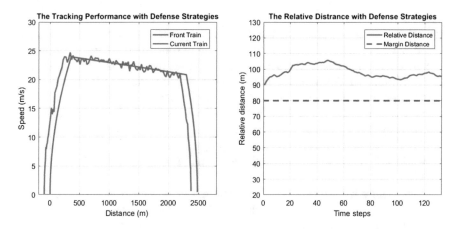

Fig. 8.11 The tracking performance of the train with defense strategies

To present more insights about our secure design, we also show the dynamics of the cyber state θ in different cases, illustrated in Fig. 8.12. In the jamming case without protection, the cyber state can increases sharply, leading to an inadequate estimation. However, under the protection of the defense mechanism, the cyber state stays at a lower level.

8.6 Conclusions and Notes

In this chapter, we have developed a mechanism to secure the CBTC system from jamming attacks, which aim to deteriorate communication performance. We have used a Markov chain to model the dynamics of the cyber state and developed a

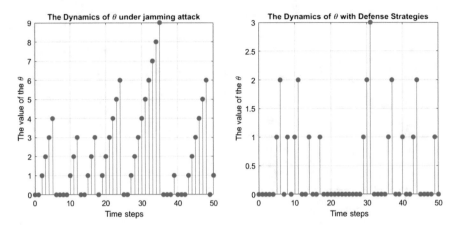

Fig. 8.12 The dynamic of the cyber state θ under the jamming attack

stochastic zero-sum game to capture the interactions between the transmitter and the jammer. The quality of communications has an impact on the control performance of CBTC systems. The stochastic game model takes into account this impact, and its equilibrium solution yields an impact-aware security strategy. We have applied a dynamic programming approach to find the equilibrium and completely characterize it in a two-channel case. In the experiments, we have evaluated the improvement of the performance in the cyber and physical layers with the proposed secure mechanism. Results have corroborated the properties of the value functions and have shown that the impact-aware strategies have significantly improved the security of CBTC.

This chapter has developed a finite-state discrete-time stochastic game framework to model the dynamic interactions between a jammer and a CBTC communication system. The stochastic games have been widely used in cybersecurity, including network configurations [200, 208], wireless security [6, 207], cyber-physical systems [107], and cyber deception [61]. In [200], stochastic games are used to develop optimal configuration policies for intrusion detection systems. The dynamic game formulation naturally leads to a minimax Q-learning algorithm that can learn equilibrium policies without prior knowledge of the game parameters. One important extension of stochastic games is to the ones under partial information. For example, the one-side partially observable stochastic games have been successfully used to capture the scenarios where one player can perfectly observe the state while the other player can only observe the state through partial observations. This class of games is particularly useful for cyber deception where the one-sided, partial observation naturally captures the information asymmetry between the defender and the attacker. See [60, 61] for more details. The one-sided partially observable games can be naturally extended to two-sided partial observations. However, the difficulty lies in hierarchical beliefs. One popular approach is to assume that the players have common priors à la Harsanyi [56, 57], and the analysis of the games dispense with hierarchical beliefs and hence become tractable.

Chapter 9
Secure Estimation of CPS with a Digital Twin

9.1 Using Digital Twin to Enhance Security Level in CPS

Due to the increasing number of security-related incidents in CPSs, many researchers have studied the features of these attacks and developed relevant defense strategies. Among many attack models, we focus on data-integrity attacks, where the attackers modify the original data used by the system or inject unauthorized data to the system [110]. The data-integrity attacks can cause catastrophic consequences in CPSs. For instance, the fake data may deviate the system to a dangerous trajectory or make the system oscillate with a significant amplitude, destabilizing the system. Therefore, how to mitigate the impact of data-integrity attacks becomes a critical issue in the security design of CPSs.

One typical data-integrity attack for CPSs is the sensor-and-estimation (SE) attack, where the attackers tamper the sensing or estimated information of the CPSs [123]. Given the SE attack, Fawzi et al. [48] have studied a SE attack and proposed algorithms to reconstruct the system state when the attackers corrupt less than half of the sensors or actuators. Pajic et al. [120] have extended the attack-resilient state estimation for noisy dynamical systems. Based on Kalman filter, Chang et al. [27] have extended the secure estimation of CPSs to a scenario in which the set of attacked nodes can change over time. However, to recover the estimation, the above work requires a certain number of uncorrupted sensors or a sufficiently long time. Those approach es might introduce a non-negligible computational overhead, which is not suitable for time-critical CPSs, e.g., real-time systems. Besides, since all the sensors do not have any protection, the attacker might easily compromise a large number of sensors, violating the assumptions of the above work.

Instead of recovering the estimation from SE attacks, researchers and operators also focus on attack detection [44]. However, detecting a SE attack could be challenging since the attackers' strategies become increasingly sophisticated. Using

Q. Zhu, Z. Xu, *Cross-Layer Design for Secure and Resilient Cyber-Physical Systems*, Advances in Information Security 81, https://doi.org/10.1007/978-3-030-60251-2_9

the conventional statistic detection theory, e.g., Chi-square detection, may fail to discover an estimation attack if the attackers launch a stealthy attack [111]. Chen et al. [34] have developed optimal attack strategies, which can deviate the CPSs subject to detection constraints. Hence, the traditional detection theory may not sufficiently address the stealthy attacks in which the attackers can acquire the defense information. Besides, using classical cryptography to protect CPSs will introduce significant overhead, degrading the control performance of delay-sensitive CPSs [178].

The development of Digital Twin (DT) provides essential resources and environments for detecting sophisticated attacks. A DT could be a virtual prototype of a CPS, reflecting partial or entire physical processes [99]. Based on the concept of DT, researchers have developed numerous applications for CPSs [163]. The most relevant application to this chapter is using a DT to achieve fault detection [18]. Similarly, we use the DT to monitor the estimation process, mitigating the influence of the SE attack.

In this chapter, we focus on a stealthy estimation attack, where the attackers know the defense strategies and aim to tamper the estimation results without being detected. Figure 9.1 illustrates the underlying architecture of the proposed framework. To withstand the attack, we design a Chi-square detector, running in a DT. The DT connects to a group of protected sensing devices, collecting relevant evidence. We use cryptography (e.g., digital signature or message authentication code) to preserve the evidence from the attack. Hence, the DT can use the evidence to monitor the estimation of the physical systems. The cryptographic overhead will not affect physical performance since the execution of the real plant does not depend on the DT.

Different from the work [27, 48, 120], we have designed two independent channels, i.e., one is protected by standard cryptography, and the other one is the general sensing channel. Figure 9.2 illustrates the structure of the framework. The main advantage of this structure is the cryptographical overhead will affect the

Fig. 9.1 The Stealthy Estimation Attack and the Defense Mechanism based on a Digital Twin: the attacker aims to modify the estimation results in a CPS, while the Digital Twin protects the system by monitoring the results

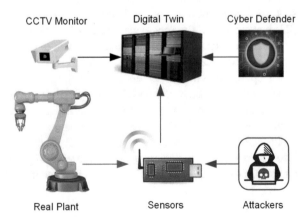

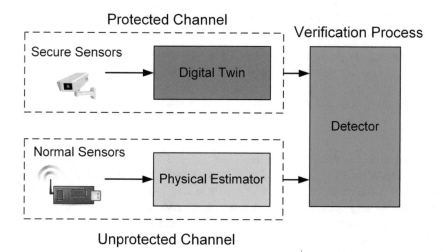

Fig. 9.2 The Architecture of the Framework: the Digital Twin has two channels to obtain the sensing information of the plant: one is secure, and the other is not secure. The secure channel provides less accurate data with a heavy computational overhead. DT only uses the secure channel to run the verification, and the computational overhead has negligible impact on the physical performance since these two channels are independent

control performance of the physical system due to the independency between these two channels.

To analyze whether a stealthy attack can bypass DT's detector, we use game theory to find the optimal attack and defense strategies. Game theory has been an essential tool in designing security algorithms since we can use it to search for the optimal defense strategies against intelligent attackers [101]. One related game model that can capture the detection issue is the Signaling Game (SG) [151]. However, instead of using the SG, we use a Signaling Game with Evidence (SGE), presented in [128], to protect the system from the attack. In an SGE, the DT's detector will provide critical evidence to explore the stealthy attack. After integrating DT's detector with CPSs, we use an SGE to study the stealthy attack and develop optimal defense strategies based on the equilibria of the SGE. Our analytical results show that the stealthy attackers have to sacrifice the impact on the physical system to avoid detection.

9.2   System Modelling and Characterization

In this section, we first introduce the dynamic model of a CPS. Secondly, we define a stealthy estimation attack model. Based on a Digital Twin (DT), we design a Chi-square detector to monitor the estimation process. Finally, we define a Signaling Game with Evidence (SGE) to characterize the features of the stealthy attack.

9.2.1   System Model and Control Problem of a CPS

Suppose that the physical layer of a CPS is a control system. We assume that the control system can be linearized as a linear discrete-time system, given by

$$x_{k+1} = Ax_k + Bu_k + w_k, \tag{9.1}$$

$$y_k = Cx_k + v_k, \tag{9.2}$$

where $k \in \mathbb{Z}_+$ is the discrete-time instant; $x_k \in \mathbb{R}^{n_x}$ is the system state with an initial condition $x_0 \sim \mathcal{N}(0_{n_x}, \Sigma_x)$, and $\Sigma_x \in \mathbb{R}^{n_x \times n_x}$ is the covariance matrix; $u_k \in \mathbb{R}^{n_u}$ is the control input; $y_k \in \mathbb{R}^{n_y}$ is the sensor output; $w_k \in \mathbb{R}^{n_x}$ and $v_k \in \mathbb{R}^{n_y}$ are additive zero-mean Gaussian noises with covariance matrices Σ_w and Σ_v with proper dimensions; A, B, and C are constant matrices with proper dimensions.

Given system model (9.1), we design a control policy $\mu : \mathbb{R}^{n_x} \rightarrow \mathbb{R}^{n_u}$ by minimizing the following expected linear quadratic cost function, i.e.,

$$J_{\text{LQG}} = \lim_{N \rightarrow \infty} \sup \mathbb{E}\left\{ \frac{1}{N} \sum_{k=0}^{N-1} \left(x_k^T Q x_k + u_k^T R u_k \right) \right\}, \tag{9.3}$$

where $Q \in \mathbb{R}^{n_x \times n_x}$ and $R \in \mathbb{R}^{n_u \times n_u}$ are positive-definite matrices.

Note that the controller cannot use state x_k directly, i.e., we need to design an observer to estimate x_k. Hence, minimizing function (9.3) is a Linear Regulator Gaussian (LQG) problem. According to the separation principle, we can design the controller and state estimator separately. The optimal control policy $\mu : \mathbb{R}^{n_x} \rightarrow \mathbb{R}^{n_u}$ is given by

$$\mu_k(x_k) := K_k x_k, \text{ with } K_k := -(R + B^T V_k B) B^T V_k, \tag{9.4}$$

where $V_k \in \mathbb{R}^{n_x \times n_x}$ is the solution to the linear discrete-time algebraic Riccati equation

$$V_{k+1} = Q + A^T V_k (A - B K_k), \text{ with } V_0 = I_{n_x}, \tag{9.5}$$

and $I_{n_x} \in \mathbb{R}^{n_x \times n_x}$ is an identity matrix.

We assume that (A, B) is stabilizable. Then, V_k will converge to a constant matrix V when k goes to infinity, i.e.,

$$\lim_{k \rightarrow \infty} V_k = V, \text{ and } \lim_{k \rightarrow \infty} \mu_k(x) = \mu(x) = Kx.$$

In the next subsection, we will use a Kalman filter to estimate x_k such that the controller (9.4) uses this estimated value to control the physical system.

9.2.2 Kalman Filter Problem

To use controller (9.4), we need to design an estimator. Let $\hat{x}_k \in \mathbb{R}^{n_x}$ be the estimation of x_k and $\hat{e}_k := \hat{x}_k - x_k$ be the error of the estimation. Given the observation $\mathcal{Y}_{k-1} := \{y_0, y_1, \ldots, y_{k-1}\}$, we aim to solve the following Kalman filtering problem, i.e.,

$$\min_{\hat{x}_k \in \mathbb{R}^{n_x \times n_x}} \mathbb{E}[(\hat{x}_k - x_k)^T (\hat{x}_k - x_k)|\mathcal{Y}_{k-1}]. \tag{9.6}$$

To solve (9.6), we need the following lemma, which characterizes a conditioned Gaussian distribution.

Lemma 9.1 ([122]) *If $a \in \mathbb{R}^{n_a}$, $b \in \mathbb{R}^{n_b}$ are jointly Gaussian with means $\bar{a}$, $\bar{b}$ and covariances Σ_a, Σ_b, and $\Sigma_{ab} = \Sigma_{ba}^T$, then given b, distribution a is a Gaussian with*

$$\mathbb{E}[a|b] = \bar{a} + \Sigma_{ab} \Sigma_b^{-1}(b - \bar{b}),$$

$$Cov[a|b] = \Sigma_a - \Sigma_{ab} \Sigma_b^{-1} \Sigma_{ba}.$$

To use Lemma 9.1, we define the covariance matrix of $\hat{e}_k$ as

$$\hat{P}_k := \mathbb{E}[\hat{e}_k \hat{e}_k^T | y_{k-1}] = \mathbb{E}[(\hat{x}_k - x_k)(\hat{x}_k - x_k)^T | y_{k-1}],$$

with $\hat{P}_0 = \Sigma_x$. Using the results of Lemma 9.1, we compute the optimal estimation iteratively, i.e.,

$$\hat{x}_k = A_K \hat{x}_{k-1} + \hat{L}_k(y_{k-1} - C\hat{x}_{k-1}), \tag{9.7}$$

where $A_K := (A + BK)$. Gain matrix $\hat{L}_k \in \mathbb{R}^{n_x \times n_y}$ and covariance $\hat{P}_k \in \mathbb{R}^{n_x \times n_x}$ are updated using

$$\hat{L}_k := A_K \hat{P}_k C^T (\Sigma_v + C\hat{P}_k C^T)^{-1},$$

$$\hat{P}_{k+1} := \Sigma_w + (A_K - \hat{L}_k C)\hat{P}_k A_K^T.$$

Assuming that (A, C) is detectable, we can obtain that

$$\lim_{k \to \infty} \hat{P}_k = \hat{P}^*, \tag{9.8}$$

where $\hat{P}^* \in \mathbb{R}^{n_x \times n_x}$ is a constant positive-definite matrix.

9.2.3   Stealthy Estimation Attack

CPSs face an increasing threat in recent years. Numerous attack models for CPSs or networked control systems (NCSs) have been introduced in [165]. Among those attacks, one major attack is the data-integrity attack, where the attacker can modify or forge data used in the CPSs. For example, Liu et al. [96] have studied a false-data-injection attack that can tamper the state estimation in electrical power grids.

To mitigate the impact of a data-integrity attack on the state estimation, researchers have designed a Chi-square detector to monitor the estimation process. However, Mo et al. [111] have analyzed stealthy integrity attack, which can pass the Chi-square detector by manipulating the perturbation of the injected data. Therefore, the main challenge is that the conventional fault detectors fail to protect the system from a stealthy attack. One real attack that can achieve the objective is the Advanced Persistent Threat (APT) [160], which can compromise a cyber system by executing a zero-day exploration to discover the vulnerabilities.

In our work, we consider an intelligent attacker who can launch a stealthy estimation attack to tamper the estimation results. Figure 9.3 illustrates how the attacker achieves its objective. The attacker can either modify the data in the sensors or the data in the estimator. Besides tampering the estimation results, the attacker is also aware of the intrusion detector. The attacker can know the defense strategy and play a stealthy attack to remain unknown.

In the next subsection, based on the Digital Twin (DT), we design a cyber defender to withstand the stealthy estimation attack and discuss the benefits introduced by the DT. After presenting the game model, we will discuss the optimal defense strategies explicitly in Sect. 9.3.

Fig. 9.3 The Stealthy Estimation Attack: intelligent attacker aims to deviate the state by modifying the estimation results

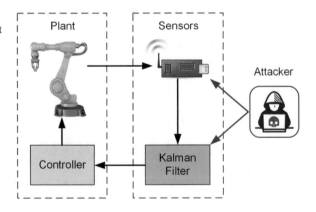

9.2.4 Digital Twin for the CPS

As mentioned above, an intelligent attacker can learn the defense strategy and launch a stealthy estimation attack, which can modify the estimation results without being detected by the conventional detector, e.g., a Chi-square detector. To resolve the issue, we aim to design an advanced detector based on a Digital Twin (DT). After that, we use a game-theoretical approach to develop an optimal defense strategy.

Given the system information, we design a DT with the following dynamics

$$\tilde{x}_k = A_K \tilde{x}_{k-1} + \tilde{L}_k(z_{k-1} - D\tilde{x}_{k-1}), \tag{9.9}$$

$$z_k = Dx_k + d_k,$$

where $\tilde{x}_k \in \mathbb{R}^{n_x}$ is the DT's estimation of state x_k; $z_k \in \mathbb{R}^{n_z}$ is the DT's observation; $D \in \mathbb{R}^{n_z \times n_x}$ is a constant matrix, d_k is a Gaussian noise with a covariance matrix $\Sigma_d \in \mathbb{R}^{n_z}$.

Similar to problem (9.6), we compute $\tilde{L}_k$ using the following iterations:

$$\tilde{L}_k = A_K \tilde{P}_k D^T (\Sigma_d + D\tilde{P}_k D^T)^{-1},$$

$$\tilde{P}_{k+1} = \Sigma_w + (A_K - \tilde{L}_k D)\tilde{P}_k A_F^T,$$

where $\tilde{P}_0 = \Sigma_x$ and $\tilde{P}_k$ is defined by

$$\hat{P}_k := \mathbb{E}[(\tilde{x}_k - x_k)(\tilde{x}_k - x_k)^T | z_{k-1}] = \mathbb{E}[\tilde{e}_k \tilde{e}_k^T | z_{k-1}].$$

where $\tilde{e}_k := \tilde{x}_k - x_k$ is the DT's estimation error. We also assume that (A_K, D) is detectable, i.e., we have

$$\lim_{k \to \infty} \tilde{P}_k = \tilde{P}^*. \tag{9.10}$$

Figure 9.4 illustrates the architecture of a CPS with a DT. We summarize the main differences between Kalman filter (9.7) and the DT's estimator (9.9) as follows. Firstly, the Kalman filter will use all available sensing information y_k to obtain estimation $\hat{x}_k$. While the DT's estimator just uses a minimum sensing information $z_k \in \mathbb{R}^{n_z}$ to predict x_k as long as (A, D) is detectable, i.e., $n_y \geq n_z$. This feature reduces the dimension of z_k, making it easier to protect z_k. Secondly, we do not require a high accuracy for z_k, since we only use z_k for attack detection. Hence, in general, $\hat{P}^*$ and $\tilde{P}^*$ satisfy the condition that $\text{tr}(\hat{P}^*) \leq \text{tr}(\tilde{P}^*)$, where $\text{tr}(P)$ is the trace of matrix P.

Thirdly, we do not use any cryptography to protect y_k since the overhead introduced by the encryption scheme will degrade the performance of the physical system. However, we can use cryptography, such as Message Authentication Code (MAC) [14] or Digital Signature (DS) [106], to protect the integrity of z_k. The overhead caused by the cryptography will not affect the physical system because

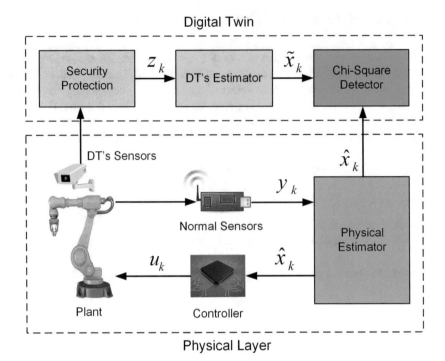

Fig. 9.4 The CPS with a DT: the DT uses a secure observation z_k to obtain an estimation $\tilde{x}_k$; given $\tilde{x}_k$, we use a Chi-square detector monitor estimation result $\hat{x}_k$

it depends on z_k. Besides, we can put the DT into a supercomputer or a cloud to resolve the overhead issue.

To sum up, z_k is an observation that is less accurate but more secure than y_k. Given the distinct features of y_k and z_k, we use y_k to estimate x_k for the physical control and use z_k for the detection in the DT.

Given DT's estimator, we construct a Chi-square detector to monitor estimation result $\hat{x}_k$ at each time k. As illustrated in Fig. 9.4, we build the detector in the DT by comparing $\tilde{x}_k$ and $\hat{x}_k$. The Chi-square detector generates a detection result $q_k \in \mathcal{Q} := \{0, 1, 2\}$ at time k, where $q_k = 0$ means the result is qualified, $q_k = 1$ means the result is unqualified, $q_k = 2$ means the result is detrimental. When $q_k = 2$, the DT should always reject the estimation and send an alarm to the operators.

To design the detector, we define $\phi_k := \tilde{x}_k - \hat{x}_k$. Since $\tilde{x}_k$ and $\hat{x}_k$ are Gaussian distributions, ϕ_k is a also Gaussian distribution with a zero-mean vector and a covariance matrix, i.e., $\phi \sim \mathcal{N}(0_{n_x}, \Sigma_\phi)$. Furthermore, we define that

$$\chi_k^2 := (\tilde{x}_k - \hat{x}_k)^T \Sigma_\phi^{-1} (\tilde{x}_k - \hat{x}_k). \tag{9.11}$$

Then, χ_k^2 follows a Chi-square distribution. We define a Chi-square detector as the following:

$$q_k = f_q(m_k) := \begin{cases} 0, & \text{if } \chi_k^2 \leq \rho_1; \\ 1, & \text{if } \chi_k^2 \in (\rho_1, \rho_2]; \\ 2, & \text{if } \chi_k^2 > \rho_2; \end{cases} \tag{9.12}$$

where ρ_1, ρ_2 are two given thresholds, and they satisfy that $\rho_2 > \rho_1 > 0$; $f_q : \mathbb{R}^{n_x} \to \mathcal{Q}$ is the detection function.

Using the above Chi-square detector, we can achieve fault detection. However, the work [111] has shown that intelligent attackers can constrain the ability of the Chi-square detector by manipulating the amount of injected data. In the following subsection, we will introduce a stealthy sensor attack that aims to remain stealthy while modifying the estimation.

9.2.5   General Setup of Signaling Game with Evidence

Due to the existence of attacks, the DT's might not be able to monitor the actual value estimation directly. Instead, the DT's can read a message provided by the estimator. According to our attack model, the attacker can compromise the estimator. Hence, the estimator can have two identities, i.e., a benign estimator or a malicious estimator. The DT aims to verify the estimator's identity by monitoring the estimation results. In Fig. 9.5, we can view DT's monitoring process as a message-sending process, i.e., the estimator sends an estimation result to the DT

Fig. 9.5 The DT's Monitoring Process: we can view the DT's monitoring process as a message-sending process, i.e., the estimator sends a message to the DT for verification

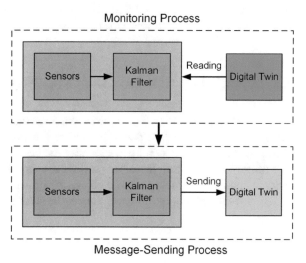

Fig. 9.6 The Architecture of
the Proposed SGE for a CPS:
the physical estimator sends
the estimation to the DT,
which uses its secure
evidence to verify the identity
of the estimator

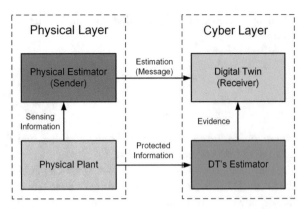

for verification. To capture the interactions between the estimator and DT, we use a
Signaling Game with Evidence (SGE), introduced in Sect. 6.5.

Figure 9.6 illustrates how an SGE captures the interactions of the physical plant
and DT in our proposed framework. In this cyber-physical SGE, we have two
players: one is the estimator, and the other one is the DT. The estimator has a
private identity, denoted by $\theta \in \Theta := \{\theta_0, \theta_1\}$, where θ_0 means the estimator
is benign, and θ_1 means the estimator is malicious. According to its identity, the
estimator will choose an estimation message $m \in \mathcal{M}$ and send it to the DT for
verification. After receiving the message, the DT should determine if the message
passes the verification. The estimator and DT have their own utility functions
$U_i : \Theta \times \mathcal{M} \times \mathcal{A} \to \mathbb{R}$, for $i \in \{s, r\}$. Detailed information about SGE has
been introduced in Sect. 6.5.

In the following sections, we will find the optimal defense strategy of the DT by
finding the PBNE. In the next section, we define the utility functions explicitly and
find the PBNE of the proposed SGE. Given the PBNE, we can identify the optimal
defense strategies. We summarize the essential notations of this chapter in Table 9.1.

9.3   Equilibrium Results of the Cyber SGE

In this section, we aim to find the optimal defense strategy against a stealthy sensor
attack. To this end, we first define the utility functions, which capture the profit
earned by the players. Secondly, we identify the best response of the players when
they observe or anticipate the other player's strategy. Finally, we present a PBNE
under the players' best response and obtain an optimal defense strategy for the DT.
We analyze the stability of the system under the stealthy attack.

Table 9.1 Summary of the notations and functions

Notation	Description
$x \in \mathbb{R}^{n_x}$	Real system state
$\hat{x} \in \mathbb{R}^{n_x}$	True estimation
$\tilde{x} \in \mathbb{R}^{n_x}$	DT's estimation
$y \in \mathbb{R}^{n_y}$	Physical observation
$z \in \mathbb{R}^{n_z}$	DT's observation
$m \in \mathcal{M}$	Sender's message
$f_q : \mathcal{M} \to \mathcal{Q}$	Detection function
$f_b : (0, 1) \times \mathcal{Q} \to (0, 1)$	Belief-update function
$\rho_1 > 0$ and $\rho_2 > 0$	Thresholds of the Chi-square detector
$\theta \in \Theta := \{\theta_0, \theta_1\}$	Sender's identity
$a \in \mathcal{A} := \{0, 1\}$	DT's action
$q \in \mathcal{Q} := \{0, 1, 2\}$	Detection result
$U_i(\theta, m, a), i \in \{s, r\}$	Utility function of player i
$\pi(\theta) \in [0, 1]$	Belief about sender's identity

9.3.1 SGE Setup for the CPSs

In this work, we use an SGE to capture the interactions between the physical estimator and the DT. In our scenario, the message set is just the estimation set, i.e., $\mathcal{M} := \mathbb{R}^{n_x}$. The DT monitors the estimation m_k and chooses an action $a \in \mathcal{A} := \{0, 1\}$. Action $a = 1$ means the estimation passes the verification, while action $a = 0$ means the verification fails, and the DT will send an alarm to the operators.

In the next step, we define the utility functions of both players, explicitly. Firstly, we define sender's utility functions $U_s(\theta, m, a)$. Since the sender has two identifies, we need to define two types of utility functions for the sender. The utility function with $\theta = 0$ is defined by

$$U_s(\theta_0, m_k, a_k) := -\mathbb{E}[(m_k - x_k)^T(m_k - x_k)], \tag{9.13}$$

In (9.13), we can see that $U_s(\theta_0, m_k, a_k)$ is independent of action a_k, and maximizing $U_s(\theta_0, m_k, a_k)$ is equivalent to the estimation problem (9.6). Hence, given (9.13), the benign sender always sends the true estimation result $\hat{x}_k$, defined by (9.7), regardless of action a_k.

For the malicious estimator, we define its utility function as

$$U_s(\theta_1, m_k, a_k) := \mathbb{E}[(m_k - x_k)^T(m_k - x_k)] \cdot \mathbf{1}_{\{a_k=1\}}, \tag{9.14}$$

where $\mathbf{1}_{\{s\}} = 1$ if statement s is true. In (9.14), we see that the motivation of the attacker is to deviate the system state as much as possible while remaining undiscovered. However, the attacker's utility will be zero if the DT detects the attack.

Secondly, we define the DT's utility function. Note that the DT's utility function should depend on the identity of the sender. When the estimator is benign, i.e., $\theta = 0$, the DT should choose $a_k = 1$ to accept the estimation. When the estimator is malicious, i.e., ($\theta = 1$), the DT should choose $a_k = 0$ to reject the estimation and send an alarm to the operators. Given the motivations, we define $U_r(\theta, \hat{x}, a)$

$$U_r(\theta_0, m_k, a_k) := -(\tilde{x}_k - m_k)^T \tilde{Q}_0 (\tilde{x}_k - m_k) \cdot \mathbf{1}_{\{a_k = 0\}},$$

$$U_r(\theta_1, m_k, a_k) := -(\tilde{x}_k - m_k)^T \tilde{Q}_1 (\tilde{x}_k - m_k) \cdot \mathbf{1}_{\{a_k = 1\}}.$$

where $\tilde{Q}_0, \tilde{Q}_1 \in \mathbb{R}^{n_x \times n_x}$ are positive-definite matrices. The weighting matrices will affect the receiver's defense strategy. A large value of $\mathrm{tr}(\tilde{Q}_1)$ will lead to a conservative strategy, while a large value of $\mathrm{tr}(\tilde{Q}_0)$ will lead a radical one. Readers can receive more details in Proposition 9.1.

In the next subsection, we analyze the behaviors of the players and obtain the best-response strategies. Note that function $U_r(\theta, m_k, a_k)$ is deterministic. The reason is that the DT can observe $\hat{x}_k$ and $\tilde{x}_k$ at time k, explicitly. However, the physical estimator cannot observe x_k at time k.

9.3.2 Best Response of the Players and a PBNE of the SGE

We first analyze the best response of the DT. Given belief $\pi_k(\theta)$, message m_k, and detection result q_k, we present the following theorem to identify DT's best response.

Proposition 9.1 *(DT's Best Response) Given $q_k = f_q(m_k)$, the DT will choose a_k according to the following policy,*

$$\sigma_r^*(a_k = 1|q_k) = \begin{cases} 1, & \text{if } q_k \neq 2, \pi_k(\theta_0) \geq \beta; \\ 0, & \text{if } q_k \neq 2, \pi_k(\theta_0) < \beta; \\ 0, & \text{if } q_k = 2; \end{cases} \qquad (9.15)$$

$$\sigma_r^*(a_k = 0|q_k) = 1 - \sigma_r^*(a_k = 1|q_k), \qquad (9.16)$$

where β is defined by

$$\beta := \frac{(\tilde{x}_k - m_k)^T \tilde{Q}_1 (\tilde{x}_k - m_k)}{(\tilde{x}_k - m_k)^T (\tilde{Q}_0 + \tilde{Q}_1)(\tilde{x}_k - m_k)}. \qquad (9.17)$$

Proof Note that

$$\mathbb{E}[U_r(\theta, m_k, a_k = 1)|q_k] \geq \mathbb{E}[U_r(\theta, m_k, a_k = 0)|q_k]$$

$$\Leftrightarrow \quad a_k = 1 \text{ if } \pi_k(\theta_0) \geq \beta,$$

where β is defined by (9.17). This completes the proof.   □

Remark 9.1 Given Proposition 9.1, we note that the DT uses a pure strategy since it can make its decision after observing detection result q_k and message m_k.

In the next step, we consider the best response of the estimator. If the estimator is benign, i.e., $\theta = \theta_0$, the optimal estimation should be (9.7). Therefore, the optimal utility of the benign estimator is given by

$$U_s(\theta_0, \hat{x}_k, a_k) = \mathbb{E}[(\hat{x}_k - x_k)^T (\hat{x}_k - x_k)] = \mathrm{tr}(\hat{P}_k),$$

where $\mathrm{tr}(P)$ is the trace of matrix P. The following theorem shows the optimal mixed strategy of the benign estimator.

Proposition 9.2 (Best Response of the Benign Estimator) *Given the DT's best response (9.15), the optimal mixed strategy of the benign estimator, i.e., $\theta = \theta_0$, is given by*

$$\sigma_s^*(f_q(m_k) = 0|\theta_0) = F_\chi(\rho_1, n_x), \tag{9.18}$$

$$\sigma_s^*(f_q(m_k) = 1|\theta_0) = F_\chi(\rho_2, n_x) - F_\chi(\rho_1, n_x), \tag{9.19}$$

$$\sigma_s^*(f_q(m_k) = 2|\theta_0) = 1 - F_\chi(\rho_2, n_x), \tag{9.20}$$

where $\hat{x}_k$ is defined by (9.7), $F_\chi(\rho, n) : \mathbb{R}_+ \to [0, 1]$ is the Cumulative Distribution Function (CDF) of the Chi-square distribution with $n \in \mathbb{Z}_+$ degrees.

Proof Note that the benign estimator will choose $m_k = \hat{x}_k$, defined by (9.7). According to definition (9.11), we know that $(\tilde{x}_k - \hat{x}_k)$ follows a Chi-square distribution with n_x degrees. Hence, we have

$$\Pr(\chi_k^2 \le \rho_1) = F_\chi(\rho_1, n_x),$$
$$\Pr(\chi_k^2 > \rho_2) = 1 - F_\chi(\rho_2, n_x),$$
$$\Pr(\chi_k^2 \in (\rho_1, \rho_2]) = F_\chi(\rho_2, n_x) - F_\chi(\rho_1, n_x).$$

Combining the above equations with Chi-square detector (9.12) yields mixed strategies (9.18)–(9.20). $\qquad\square$

Remark 9.2 Note that the benign estimator always chooses the optimal estimation 9.7. However, from DT's perspective in this game, the real mixed strategies of the benign estimator are (9.18)–(9.20) because of uncertainty introduced by the noises.

From the perspective of the malicious estimator, it needs to select $\sigma_s(q_k|\theta_1)$ such that $\pi_k(\theta_0) \ge \beta$. Given the attackers' incentive, we obtain the following theorem.

Proposition 9.3 (Best Response of the Malicious Estimator)

$$\sigma_s^*(f_q(\xi_{k,1}) = 0|\theta_1) = F_\chi(\rho_1, n_x), \tag{9.21}$$

$$\sigma_s^*(f_q(\xi_{k,2}) = 1|\theta_1) = 1 - F_\chi(\rho_1, n_x), \tag{9.22}$$

where $\xi_{k,1}, \xi_{k,2}$ are the solutions to the following problems:

$$\xi_{k,1} \in \underset{m \in \mathcal{M}_{\rho_1}(\tilde{x}_k)}{\arg\max} \quad U_s(\theta_1, m, a_k = 1), \tag{9.23}$$

$$\xi_{k,2} \in \underset{m \in \mathcal{M}_{\rho_2}(\tilde{x}_k)}{\arg\max} \quad U_s(\theta_1, m, a_k = 1), \tag{9.24}$$

with spaces $\mathcal{M}_{\rho_1}(\tilde{x}_k)$ and $\mathcal{M}_{\rho_2}(\tilde{x}_k)$ defined by

$$\mathcal{M}_{\rho_1}(\tilde{x}_k) := \left\{ m \in \mathbb{R}^{n_x} \,\middle|\, \|\tilde{x}_k - m\|^2_{\Sigma_\phi^{-1}} \leq \rho_1 \right\},$$

$$\mathcal{M}_{\rho_2}(\tilde{x}_k) := \left\{ m \in \mathbb{R}^{n_x} \,\middle|\, \|\tilde{x}_k - m\|^2_{\Sigma_\phi^{-1}} \in (\rho_1, \rho_2] \right\}.$$

Proof Firstly, the attacker has no incentive to choose $m_k \notin \mathcal{M}_{\rho_1}(\tilde{x}_k) \cup \mathcal{M}_{\rho_2}(\tilde{x}_k)$ because its utility will be zero. Secondly, the attacker aims to choose the mixed strategy $\sigma_s^*(q_k = 1|\theta_1)$ as large as possible since it can make a higher damage to the system. Then, we show that the optimal mixed strategy of the attacker is given by (9.21) and (9.22). To do this, based on (6.3), we consider the following belief update:

$$
\begin{aligned}
&\pi_{k+1}(\theta_0) \\
&= \frac{\sigma_s^*(q_k \neq 0|\theta_0)\pi_k(\theta_0)}{\sigma_s^*(q_k \neq 0|\theta_0)\pi_k(\theta_0) + \sigma_s(q_k \neq 0|\theta_1)(1 - \pi_k(\theta_0))} \\
&= \frac{\sigma_s^*(q_k \neq 0|\theta_0)\pi_k(\theta_0)}{\Delta\sigma_s(q_k \neq 0)\pi_k(\theta_0) + \sigma_s(q_k \neq 1|\theta_1)},
\end{aligned} \tag{9.25}
$$

where $\Delta\sigma_s(q_k \neq 0)$ is defined by

$$\Delta\sigma_s(q_k \neq 0) := \sigma_s^*(q_k \neq 0|\theta_0) - \sigma_s(q_k \neq 0|\theta_1).$$

Rearranging (9.25) yields that

$$\frac{\pi_{k+1}(\theta_0)}{\pi_k(\theta_0)} = \frac{\sigma_s^*(q_k \neq 0|\theta_0)\pi_k(\theta_0)}{\Delta\sigma_s(q_k \neq 0)\pi_k(\theta_0) + \sigma_s(q_k \neq 0|\theta_1)}.$$

Given that $\pi_k(\theta_0) \in (0, 1]$, we have

$$
\begin{cases}
\frac{\pi_{k+1}(\theta_0)}{\pi_k(\theta_0)} > 1, & \text{if } \Delta\sigma_s(q_k \neq 0) > 0; \\
\frac{\pi_{k+1}(\theta_0)}{\pi_k(\theta_0)} = 1, & \text{if } \Delta\sigma_s(q_k \neq 0) = 0; \\
\frac{\pi_{k+1}(\theta_0)}{\pi_k(\theta_0)} < 1, & \text{if } \Delta\sigma_s(q_k \neq 0) < 0.
\end{cases}
$$

When $\pi_k(\theta_0) \in [\beta, 1)$, the attacker has to choose $\Delta\sigma_s(q_k \neq 0) = 0$ to maintain the belief at a constant. Otherwise, the belief will decrease continuously. When the belief $\pi_k(\theta_0)$ stays lower than β, the DT will send an alert to the operators. Hence, the optimal mixed strategies of the malicious estimator are given by (9.21) and (9.22). □

Given the results of Propositions 9.1, 9.2, and 9.3, we present the following theorem to characterize the unique pooling PBNE.

Theorem 9.1 (The PBNE of the Proposed SGE) *The proposed cyber SGE has a unique pooling PBNE. At the PBNE, the optimal mixed strategies of the benign and malicious sender are presented by (9.18)–(9.20) and (9.21)–(9.22). The DT has a pure strategy defined by (9.15). At the PBNE, belief $\pi_k^*(\theta_0) \in [\beta, 1)$ is a fixed point of function f_b, i.e.,*

$$
\pi_k^*(\theta_0) = f_b(\pi_k^*(\theta_0), q_k), \text{ for } q_k \in \mathcal{Q}.
$$

Proof We first show the existence of the pooling PBNE. Suppose that both estimators use strategies (9.18)–(9.20), (9.21)–(9.22), respectively, and the DT uses (9.15). Then, no player has incentive to move since these are already the optimal strategies. Besides, for any $\theta \in \Theta$, $q_k \in \mathcal{Q}$, we note that

$$
\pi_{k+1}(\theta) = f_b(\pi_k^*(\theta), q_k) = \pi_k^*(\theta),
$$

where f_b is defined by (6.3). Hence, $\pi_k^*(\theta)$ is a fixed point of function f_b, and the belief remain at $\pi_k^*(\theta)$, which means the belief stays consistently with the optimal strategies of the sender and receiver. Hence, the proposed strategies pair (σ_s^*, σ_r^*) is a PBNE.

Secondly, we show that pooling PBNE is unique. We note that the DT and the benign estimator have no incentive to move since they already choose their best strategies. In Proposition 9.3, we already show that the attacker cannot change its mixed strategies. Otherwise, the belief cannot remain constant. Hence, pooling PBNE is unique. □

Remark 9.3 Theorem 9.1 shows that the SGE admits a unique pooling PBNE, which means that an intelligent attacker can use its stealthy strategies to avoid being detected by the DT.

In the next subsection, we will analyze the stability of the system under the stealthy attack. Besides, we will also evaluate the loss caused by the attack.

9.3.3   Estimated Loss Under the Stealthy Attack

In the previous subsection, we have shown the PBNE in which the attacker can use a stealthy strategy to pass the verification of the DT. In this subsection, we will quantify the loss under the attack. Before presenting the results, we need the following lemma.

Lemma 9.2 *Given ρ_i and $\xi_{k,i}$, for $i \in \{1, 2\}$, we have the following relationship:*

$$\rho_i \lambda_{\max}(\Sigma_\phi) \geq (\tilde{x}_k - \xi_{k,i})^T (\tilde{x}_k - \xi_{k,i}), \text{ for } i = 1, 2,$$

where $\lambda_{\max}(\Sigma)$ is the greatest eigenvalue of matrix Σ.

Proof Firstly, we note that U_s is strictly convex in m_k. The solution to problem (9.23) and (9.24) must stay at the boundary. Hence, we have

$$\rho_i = (\tilde{x}_k - \xi_{k,i})^T \Sigma_\phi^{-1} (\tilde{x}_k - \xi_{k,i})$$

$$\geq \frac{(\tilde{x}_k - \xi_{k,i})^T (\tilde{x}_k - \xi_{k,i})}{\lambda_{\max}(\Sigma_\phi)} \tag{9.26}$$

Rearranging (9.26) yields that

$$\rho_i \lambda_{\max}(\Sigma_\phi) \geq (\tilde{x}_k - \xi_{k,i})^T (\tilde{x}_k - \xi_{k,i}).$$

This completes the proof.   □

Considering different estimators, we define two physical cost functions J_0 and J_1, i.e.,

$$J_0 := \lim_{N \to \infty} \mathbb{E} \left\{ \frac{1}{N} \sum_{k=0}^{N-1} \left[x_k^T Q x_k + \mu^T(m_k) R \mu(m_k) \right] \Big| \theta_0 \right\},$$

$$J_1 := \lim_{N \to \infty} \mathbb{E} \left\{ \frac{1}{N} \sum_{k=0}^{N-1} \left[x_k^T Q x_k + \mu^T(m_k) R \mu(m_k) \right] \Big| \theta_1 \right\}.$$

We define a loss function $\Delta J := J_1 - J_0$ to quantify the loss caused by the stealthy sensor attack. Given pooling PBNE defined by Theorem 9.1, we provide an upper-bound of ΔJ in the following theorem.

Theorem 9.2 (Bounded Loss) *The proposed framework can guarantee stability of the CPSs, and the value of function ΔJ is bounded by a constant, i.e.,*

$$\Delta J = J_1 - J_0$$

$$\leq \alpha_1 tr(\tilde{P}^*) - \alpha_0 tr(\hat{P}^*) + \alpha_1 \rho_1 \lambda_{\max}(\Sigma_\phi) F_\chi(\rho_1, n_x)$$

$$+ \alpha_1 \rho_2 \lambda_{\max}(\Sigma_\phi)(1 - F_\chi(\rho_1, n_x)), \tag{9.27}$$

where α_0 and α_1 are defined by

$$\alpha_0 := \frac{\lambda_{\min}(R_K)(\lambda_{\min}(G) - \lambda_{\min}(R_K))}{\lambda_{\min}(G)},$$

$$\alpha_1 := \frac{\lambda_{\max}(R_K)(\lambda_{\max}(R_K) + 2\lambda_{\max}(G))}{\lambda_{\max}(G)},$$

and $R_K := K^T R K, G := Q + R_K$; $\lambda_{\max}(W)$ and $\lambda_{\min}(W)$ are the greatest and smallest eigenvalues of matrix W.

Proof Firstly, we note that

$$\mathbb{E}\left[\|x_k\|_Q^2 + \|K\hat{x}_k\|_R^2\right]$$

$$= \mathbb{E}\left[\|x_k\|_G^2 + 2x_k^T R_K \hat{e}_k + \|\hat{e}_k\|_{R_K}^2\right]$$

$$\geq \mathbb{E}\left[\left\|\sqrt{\lambda_{\min}(G)}x_k + \frac{\lambda_{\min}(R_K)}{\sqrt{\lambda_{\min}(G)}}\hat{e}_k\right\|^2\right.$$

$$\left. + \frac{\lambda_{\min}(R_K)(\lambda_{\min}(G) - \lambda_{\min}(R_K))}{\lambda_{\min}(G)}\|\hat{e}_k\|^2\right] \geq \alpha_0 \mathrm{tr}(\hat{P}_k),$$

Using the above inequality, we observe that

$$J_0 = \lim_{N\to\infty} \frac{1}{N} \sum_{k=0}^{N-1} \mathbb{E}\left[\|x_k\|_Q^2 + \|K\hat{x}_k\|_R^2\right]$$

$$\geq \lim_{N\to\infty} \frac{1}{N} \sum_{k=0}^{N-1} \alpha_0 \mathrm{tr}(\hat{P}_k) = \alpha_0 \mathrm{tr}(\hat{P}^*), \tag{9.28}$$

where $\hat{P}^*$ is defined by (9.8). Secondly, we also note that

$$\mathbb{E}\left[x_k^T Q x_k + \mu^T(\xi_{k,i}) R \mu(\xi_{k,i})\right]$$

$$= \mathbb{E}\left[\|x_k\|_Q^2 + \|x_k + \tilde{e}_k + \xi_{k,i} - \tilde{x}_k\|_{R_K}^2\right]$$

$$\leq \mathbb{E}\left[\|x_k\|_G^2 + \left(2x_k^T R_K \tilde{e}_k + 2x_k^T R_K (\xi_{k,i} - \tilde{x}_k)\right)\right.$$

$$\left. + 2\tilde{e}_k^T R_K (\xi_{k,i} - \tilde{x}_k) + \|\xi_{k,i} - \tilde{x}_k\|_{R_K}^2 + \|\tilde{e}_k\|_{R_K}^2\right]$$

$$\leq \mathbb{E}\left[3\|x_k\|_G^2 + \frac{\lambda_{\max}^2(R_K)}{\lambda_{\max}(G)}\left(\|\tilde{e}_k\|^2 + \|\xi_{k,i} - \tilde{x}_k\|^2\right)\right.$$

$$\left. + 2\lambda_{\max}(R_K)\|\tilde{e}_k\|^2 + 2\lambda_{\max}(R_k)\|\xi_{k,i} - \tilde{x}_k\|^2\right]$$

$$\leq 3\|x_k\|_G^2 + \alpha_1 \text{tr}(\tilde{P}_k) + \alpha_1 \rho_i \lambda_{\max}(\Sigma_\phi), \tag{9.29}$$

We complete the squares to deduce the second inequality of (9.29) Similarly, we have

$$J_1 = \lim_{N\to\infty} \frac{1}{N} \sum_{k=0}^{N-1} \left\{ F_\chi(\rho_1, n_x)\mathbb{E}\left[\|x_k\|_Q^2 + \|\xi_{k,1}\|_{R_K}^2\right]\right.$$

$$\left. + \left(1 - F_\chi(\rho_1, n_x)\right)\mathbb{E}\left[\|x_k\|_Q^2 + \|\xi_{k,1}\|_{R_K}^2\right]\right\}$$

$$\leq \underbrace{\lim_{N\to\infty} \frac{3}{N} \sum_{k=0}^{N-1} \|x_k\|_G^2}_{=0} + \alpha_1 \rho_1 \lambda_{\max}(\Sigma_\phi) F_\chi(\rho_1, n_x)$$

$$+ \alpha_1 \rho_2 \lambda_{\max}(\Sigma_\phi)(1 - F_\chi(\rho_1, n_x)) + \alpha_1 \text{tr}(\tilde{P}^*), \tag{9.30}$$

where $\tilde{P}^*$ is defined by (9.10). Combining inequalities (9.28) and (9.30) yields inequality (9.27). Hence, the system is stable, and the impact of the attack is bounded by a constant.   □

Remark 9.4 Theorem 9.2 shows that the difference between J_0 and J_1 is bounded; i.e., the stealthy estimation attack cannot deviate the system to an arbitrary point even if the attacker has an infinite amount of time.

In the next subsection, we will use an application to evaluate the performance of the proposed defense strategies.

9.4   Simulation Results

In this section, we use a two-link Robotic Manipulator (RM) to investigate the impact of the estimation attacks. In the experiments, we use different case studies to analyze the performance of the proposed defense framework.

Fig. 9.7 The dynamic model
of a two-link robotic
manipulator (RM): the RM
has two links and moves in a
two-dimensional space

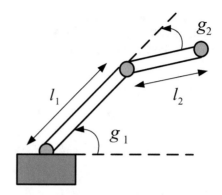

Table 9.2 Parameters of the
robotic manipulator

Parameter	Description	Value
l_1	Length of link 1	0.6 m
l_2	Length of link 2	0.4 m
r_1	Half length of link 1	0.3 m
r_2	Half length of link 2	0.2 m
η_1	Mass of link 1	6.0 kg
η_2	Mass of link 2	4.0 kg
I_1	Inertia of link 1 on z-axis	1 kg m^2
I_2	Inertia of link 2 on z-axis	1 kg m^2

9.4.1 Experimental Setup

Figure 9.7 illustrates the physical structure of the two-link RM. Variables g_1 and g_2
are the angular positions of Links 1 and 2. We summarize the parameters of the RM
in Table 9.2.

Let $g = [g_1, g_2]^T$ be the angular vector and $\tau = [\tau_1, \tau_2]^T$ be the torque input.
According to the Euler-Lagrange Equation, we obtain the dynamics of the two-link
RM as

$$M(g)\ddot{g} + S(g, \dot{g})\dot{g} = \tau, \tag{9.31}$$

where matrices $M(g)$ and $S(g, \dot{g})$ are defined by

$$M(g) := \begin{bmatrix} a + b\cos(g_2) & \delta + b\cos(g_2) \\ \delta + \cos(g_2) & \delta \end{bmatrix},$$

$$S(g, \dot{g}) := \begin{bmatrix} -b\sin(g_2)\dot{g}_2 & -b\sin(g_2)(\dot{g}_1 + \dot{g}_2) \\ b\sin(g_2)\dot{g}_1 & 0 \end{bmatrix},$$

$$a := I_1 + I_2 + \eta_1 r_1^2 + \eta_2(l_1^2 + r_2^2),$$

$$b := \eta_2 l_1 r_2, \qquad \delta := I_2 + \eta_2 r_2^2.$$

To control the two-link RM, we let τ be $\tau := M(g)a_g + S(g, \dot{g})\dot{g}$, where $a_q \in \mathbb{R}^2$ is the acceleration that we need to design. Note that $M(g)$ is positive-definite, i.e., $M(g)$ is invertible. Hence, substituting τ into (9.31) yields that

$$M(g)\ddot{g} = M(g)a_g \Rightarrow \ddot{g} = a_g.$$

Let $p \in \mathbb{R}^2$ be the position of RM's end-effector. We have

$$\ddot{p} = H(g)\ddot{g} + \dot{H}(g)\dot{g} = H(g)a_g + \dot{H}(g)\dot{g}, \tag{9.32}$$

where $H(g)$ is the Jacobian matrix. Then, we substitute $a_g := H^{-1}(g)(u - \dot{H}(g))$ into (9.32), arriving at $\ddot{p} = u$. Let $x = [p^T, \dot{p}^T]^T$ be the continuous-time state. Then, we obtain a continuous-time linear system $\dot{x} = A_c x + B_c u$. Given a sampling time $\Delta T > 0$, we discretize the continuous-time system to obtain system model (9.1). We let y_k and z_k be

$$y_k = x_k + v_k, \quad z_k = p(k\Delta T) + d_k \tag{9.33}$$

We assume that the DT uses security-protected cameras to identify the position of the end-effector.

In the experiments, we let the RM to draw a half circle on a two-dimensional space. The critical parameters are given by

$$\beta = 0.65, \ n_x = 4, \ \rho_1 = 9.49, \ \rho_2 = 18.47,$$

$$F_\chi(\rho_1, n_x) = 0.95, \quad F_\chi(\rho_2, n_x) = 0.999.$$

We have three case studies: a no-attack case, a normal-attack case, and a stealthy-attack case. In the normal-attack case, the attacker is not aware of the defense strategies and deviate the system from the trajectory, directly. In the last case, the attacker aims to tamper the estimation without being detected.

Figure 9.8, 9.9, and 9.10 illustrate the simulation results of the case studies. In Fig. 9.8a, we can see that the RM can track the trajectory smoothly when there is no attack. However, we note that DT's estimation is worse than the physical estimation, which coincides with our expectations. Figure 9.8b and c shows the value of the Chi-square and the belief of the DT. In the no-attack case, the Chi-square detector will remain silent with a low false alarm rate, and the belief stays at a high level. In Fig. 9.9a, the attackers deviate the system without considering the detection. Even though DT's estimation is not accurate, the attacker cannot tamper that. Therefore, the detector will rapidly locate the attack and send alarms to the operators. The belief of θ_0 will remain at the bottom line. In Fig. 9.10, differently, the stealthy attackers know the defense strategies and try to maintain the Chi-square value below threshold ρ_1. However, the behavior mitigates the impact of the attack, which also coincides with the result of Theorem 9.2.

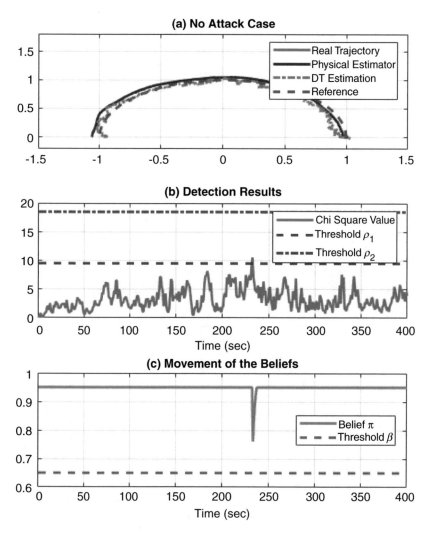

Fig. 9.8 No-attack case: (**a**) the system trajectory, physical estimation and DT's estimation; (**b**) the Chi-square value; (**c**) DT's belief $\pi(\theta_0)$

Figure 9.11 illustrates the Mean Square Error (MSE) of different cases. Figure 9.11a presents that the MSE of the physical estimator is much smaller than the DT's estimator, i.e., the physical estimator can provide more accurate sensing information. However, in Fig. 9.11b, we can see that the attacker can deviate the physical estimation to a significant MSE. Besides, under the DT's supervision, the stealthy attacker fails to generate a large MSE. The above results show that the proposed defense mechanism succeeds in mitigating the stealthy attacker's impact.

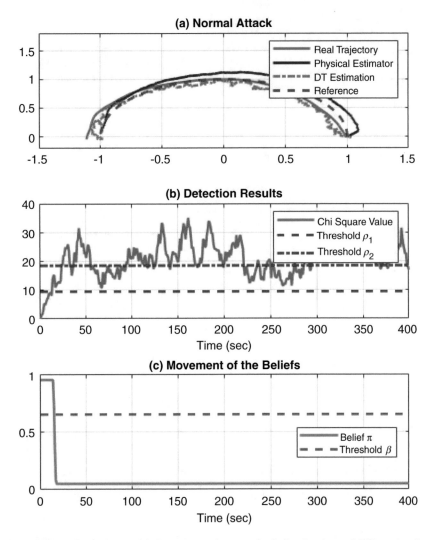

Fig. 9.9 Normal-attack case: (**a**) the system trajectory, physical estimation and DT's estimation; (**b**) the Chi-square value; (**c**) DT's belief $\pi(\theta_0)$

9.5   Conclusions and Notes

In this chapter, we have considered a stealthy estimation attack, where an attack can modify the estimation results to deviate the system without being detected. To mitigate the impact of the attack on physical performance, we have developed a Chi-square detector, running in a Digital Twin (DT). The Chi-square detector can collect DT's observations and the physical estimation to verify the estimator's identity. We have used a Signaling Game with Evidence (SGE) to study the optimal attack and

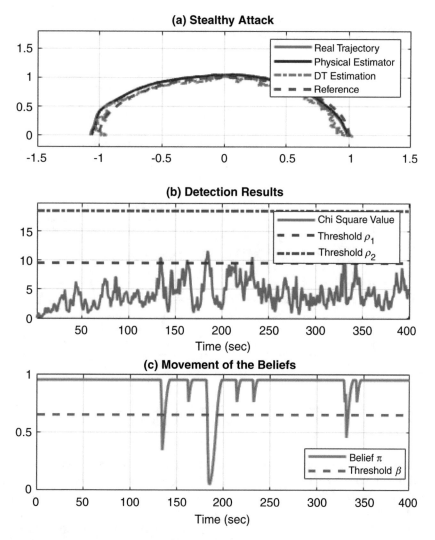

Fig. 9.10 Stealthy-attack case: (**a**) the system trajectory, physical estimation and DT's estimation; (**b**) the Chi-square value; (**c**) DT's belief $\pi(\theta_0)$

defense strategies. Our analytical results have shown that the proposed framework can constrain the attackers' ability and guarantee stability.

This chapter has leveraged signaling game models as one of the building blocks to design defense strategies to capture the sender's unknown identity. The Chi-square detection has provided evidence to the signal game framework. This type of game framework is called a signaling game with evidence. Signaling games have been used to investigate many security problems in which a sender has access to the information while a receiver cannot. This situation naturally applies to man-in-

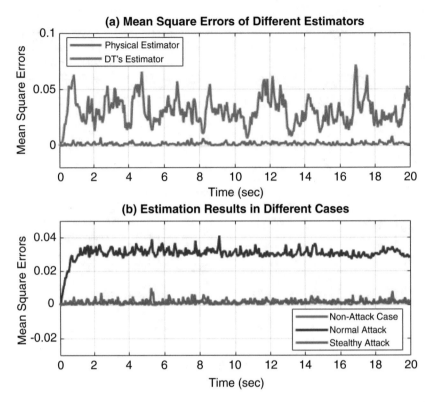

Fig. 9.11 The comparison of the Mean Square Error (MSE): (**a**) the comparison between the MSE of physical estimation and DT's estimation; (**b**) the MSE of different case studies

the-middle attack in remotely controlled robotic systems [175], spoofing attacks [191], compliance control [24, 24], deception over social networks [112], and denial of service [100, 124]. Signaling games provide a mechanism to design trust mechanisms on the received messages and the decisions to accept or reject inputs in many applications. As demonstrated in this chapter, they can be easily integrated into system designs. This design philosophy has also been applied to cloud-enabled control systems as in [125–127], where the equilibrium solutions are used as a trust management tool to select trustworthy inputs for the feedback system.

Chapter 10
Introduction to Partially Observed MDPs

10.1 Preliminaries of POMDPs

A POMDP is a controlled HMM, which consists of an S-state Markov chain $\{s_k\}$ observed via a noisy observation process z_k. Figure 10.1 illustrates the architecture of the POMDP, where action a_k affects the state and observation (detection) process of the HMM. The HMM filter outputs a posterior belief θ_k of state s_k. The posterior belief is called the *belief state*. In a POMDP, the controller (decision-maker) choose the next action a_k based on the belief θ_k.

10.1.1 Definition of a POMDP

After a brief introduction, we will present a formal definition of a POMDP. A POMDP model with an infinite horizon is a 7-tuple

$$(\mathcal{S}, \ \mathcal{A}, \ \mathcal{Z}, \ \Theta, \ \Lambda(a), \ \Phi(z), \ g(s,a)), \tag{10.1}$$

where

- $\mathcal{S} := \{1, 2, \ldots, S\}$ denotes the state space, and $s_k \in \mathcal{S}$ denotes the state of a controlled Markov chain at time $k \in \mathbb{Z}_+$.
- $\mathcal{A} := \{1, 2, \ldots, A\}$ denotes action space, and $a_k \in \mathcal{A}$ denotes the action chosen at time k by the controller.
- $\mathcal{Z} := \{1, 2, \ldots, Z\}$ denotes the observation space, and $z_k \in \mathcal{Z}$ denotes the observation record at time k.
- $\Theta := \{\theta \in \mathbb{R}^S : \sum_{i=1}^{S} \pi(i) = 1, 0 \leq \pi(i) \leq 1, \forall i \in \mathcal{S}\}$ denotes the belief space, $\theta_k \in \Theta$ denotes the posterior belief of state s_k at time k.

Q. Zhu, Z. Xu, *Cross-Layer Design for Secure and Resilient Cyber-Physical Systems*,
Advances in Information Security 81, https://doi.org/10.1007/978-3-030-60251-2_10

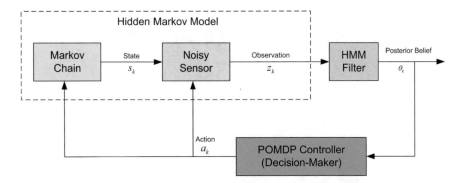

Fig. 10.1 Partially observed Markov Decision Process (POMDP) schematic setup. The Markov system together with noisy sensor constitutes a hidden Markov model (HMM). The HMM filter computes the posterior belief θ_k of the state of Markov chain. The POMDP controller (decision-maker) then choose the action a_k at time k based on belief θ_k

- $\Lambda(a) \in \mathbb{R}^{S \times S}$ denotes the transition probability matrix with elements

$$\lambda_{ij}(a) := \Pr(s_{k+1} = j | s_k = i, a_k = a), \quad \forall i, j \in \mathcal{S}.$$

- For each observation $z \in \mathcal{Z}$, $\Phi(z)$ denotes the observation distribution matrix with elements

$$\phi_{ii}(z) = \Pr(z_k = z | s_k = i), \ \phi_{ij}(z) = 0, \ \forall i, j \in \mathcal{S}, \ z \in \mathcal{Z}, \ i \neq j.$$

- Given state s_k and action a_k, the decision-maker incurs a cost $g(s_k, a_k)$.

The POMDP, defined by (10.1), is a partially observed model since the decision-maker cannot observe state s_k directly. Instead, it only observes a noisy observation z_k that depend on the state specified by the observation distribution.

To specify a POMDP problem, we need to consider a performance criterion or objective function. This chapter considers an infinite horizon objective, given by

$$J(\theta_0, \mu) = \mathbb{E}_\mu \left\{ \sum_{k=0}^{\infty} \beta^k g(s_k, \mu) \Big| \theta_0 \right\}, \tag{10.2}$$

where $\theta_0 := \Pr(s_0)$ is an initial distribution of the state, and $\beta \in (0, 1)$ is a discounted factor. $\mathbb{E}_\mu$ denotes expectation with respect to the joint probability distribution of $(s_0, z_0, x_1, z_1, \ldots,)$. The objective of the decision-maker is to determine the optimal policy sequence

$$\mu^* \in \arg \min_\mu J(\theta_0, \mu), \quad \text{for any initial prior } \theta_0 \in \Theta.$$

In the next subsection, we will define the belief state θ as well as the dynamical movement of θ

10.1.2 Belief State Formulation of a POMDP

Note that at time k, the decision maker has an information set i_k, defined by

$$i_k := \{\theta_0, a_0, z_1, \ldots, \theta_k - 1, a_{k-1}, z_k\}. \tag{10.3}$$

Given information set i_k, the decision-maker will choose its optimal action using $a_k = \mu^*(i_k)$, where μ^* is the optimal policy of the decision-maker.

Since the dimension of i_k grows with time k, we need to develop a sufficient statistic that does not grow in dimension. One way to obtain a sufficient statistic for i_k is to introduce the posterior distribution computed via the HMM filter. Hence, we define the posterior distribution of the Markov chain as

$$\theta_k(i) := \Pr(s_k = i | i_k), \quad i \in \mathcal{S}.$$

We will call the S-dimensional probability vector $\theta_k := [\theta_k(1), \ldots, \theta_k(S)]^T$ as belief state or information state at time k. We use the Bayes' rule to update the belief. Given Bayes's rule, we define a belief update function $T : \Theta \times \mathcal{Z} \times \mathcal{A} \to \Theta$, i.e.,

$$\theta_{k+1} = T(\theta_k, z_{k+1} a_k) := \frac{\Phi(z) \Lambda^T(a) \theta_k}{\sigma(\theta_k, z_{k+1}, a_k)}$$

$$\sigma(\theta_k, z_{k+1}, a_k) := \mathbf{1}_S^T \Phi(z) \Lambda^T(a) \theta.$$

Given the belief state, the decision-maker can choose its optimal action using $a_k = \mu^*(\theta_k)$, where the dimension of θ is a constant. To obtain a policy based on the belief state, we reformulate the POMDP objective (10.2) in terms of belief state. Given (10.2), we observe that

$$J(\theta_0, \mu) := \mathbb{E}_\mu \left\{ \sum_{k=0}^{\infty} \beta^k g(s_k, \mu) \Big| \theta_0 \right\}$$

$$\overset{(a)}{=} \mathbb{E}_\mu \left\{ \sum_{k=0}^{\infty} \beta^k \mathbb{E}\{g(s_k, \mu) | i_k\} \Big| \theta_0 \right\}$$

$$\overset{(b)}{=} \mathbb{E}_\mu \left\{ \sum_{k=0}^{\infty} \beta^k g^T(\mu) \theta_k \Big| \theta_0 \right\}, \tag{10.4}$$

where $g(a) := [g(1,a), \ldots, g(S,a)]^T$; equality (a) uses the smoothing property of conditional expectation [82]; equality (b) comes from the definition of expectation. Then, we aim to find the optimal policy μ^* to minimize cost function (10.4).

To find the optimal policy μ^*, we will introduce stochastic dynamic programming in the next subsection.

10.1.3   Stochastic Dynamic Programming

Since a POMDP is a continuous-state MDP with state space being the unit simplex, we can write down the dynamic programming equation for the optimal policy. Given a policy μ, we define a value function $V_\mu : \Theta \to \mathbb{R}$ as

$$V_\mu(\theta_k) := \mathbb{E}_\mu\left\{\sum_{j=k}^{\infty} \beta^{j-k} g^T(\mu)\theta_j \middle| \theta_k\right\}$$

$$= g^T(\mu)\theta_k + \beta \sum_{z\in\mathcal{Z}} V_\mu(T(\theta_k, z, \mu))\sigma(\theta_k, z, \mu), \qquad (10.5)$$

where the second equality comes from the Bellman equation. Hence, we can find the optimal policy using

$$\mu^*(\theta_k) \in \arg\min_{\mu\in\mathcal{A}}\left\{g^T(\mu)\theta_k + \beta \sum_{z\in\mathcal{Z}} V_\mu(T(\theta_k, z, \mu))\sigma(\theta_k, z, \mu)\right\}.$$

However, since the belief space Θ is uncountable, the above dynamic programming recursion does not translate into practical solutions. We need to evaluate $V_\mu(\theta)$ at each $\theta \in \Theta$, where Θ is uncountable. Hence, in the following subsection, we will introduce a method to address this issue.

10.2   Algorithms for Infinite Horizon POMDPs

As stated in the previous section, it is challenging to solve the POMDP using dynamic programming since the belief space is uncountable. In this section, we first show a piecewise linear property of the POMDPs. Then, we define a Markov Partition (MP), with which we can simplify the POMDPs. In the end, we can use Policy Iteration (PI) to solve the simplified POMDPs.

10.2.1 Piecewise Linear Property of POMDPs

Despite the belief state Θ being continuum, we present the following remarkable results based on [155]. They have shown that Bellman's equation (10.5) has a finite-dimensional characterization when the observation space $\mathcal{Z}$ is finite. The following theorem characterizes these results.

Theorem 10.1 *Consider the POMDP model (10.1) with finite action space $\mathcal{A} = \{0, 1, dots, A\}$ and finite observation space $\mathcal{Z} = \{0, 1, dots, Z\}$. At each time k, the value function $V_\mu(\theta)$ of Bellman's equation (10.5) and associated optimal policy $\mu^*(\theta)$ have the following finite dimensional characterization: $V_\mu(\theta)$ is piecewise linear and concave with respect to $\theta \in \Theta$, i.e.,*

$$V_\mu(\theta) = \eta^T(\theta, \mu) \times \theta,$$

where $\eta : \Theta \times \mathcal{A} \to \mathbb{R}^S$ satisfying

$$\eta(\theta, \mu) = g(\mu) + \beta \sum_{z \in \mathcal{Z}} \Lambda(\mu)\Phi(z)\eta(\theta', \mu). \tag{10.6}$$

Theorem 10.1 shows a very nice property for the POMDP, but the problem is still challenging to solve since the number of linear equations (10.6) is uncountable. In the next section, we will present the definition of a Markov Partition (MP), which can simplify the POMDP to a finite-space case.

10.2.2 Algorithms Based on Markov Partition

In this subsection, we aim to simplify the POMDP. One critical approach is to use the Markov Partition. Given a stationary policy μ, we will partition the belief space Θ into finite intervals. If the intervals satisfy the properties of Markov partition (MP), then, we can treat the POMDP as a discrete MDP with a finite number of states. We first present the definition of MP as follows.

Definition 10.1 (Markov Partition) Given a stationary policy μ, a partition $\mathcal{P} = \{\mathcal{P}_1, \mathcal{P}_2, \ldots, \mathcal{P}_Y\}$ of Θ is said to be Markov if for any $y \in \mathcal{Y} = \{1, 2, \ldots, Y\}$:

1. The policy μ maps all the beliefs $\theta \in \mathcal{P}_l$ to the same action, i.e., for $\theta_1, \theta_2 \in \mathcal{P}_y$, we have $\mu(\theta_1) = \mu(\theta_2)$.

2. The belief update function $T(\theta, z, \mu)$ maps all the beliefs in $\mathcal{P}_y$ to the same set $\mathcal{P}_{v(z)}$, i.e., for any $\theta_1, \theta_2 \in \mathcal{P}_y$, $T(z, \theta_1, \mu)$ and $T(z, \theta_2, \mu)$ lie in the same set $\mathcal{P}_{v(z)}$, where $v(z) \in \mathcal{Y}$.

If a POMDP admits an MP with Y number of intervals, then it behaves like an MDP. Given an MP with Y states and a policy μ, we have

$$\eta(y, \mu_y) = g(\mu_y) + \beta \sum_{z \in \mathcal{Z}} \Lambda(\mu_y) \Phi(z) \eta(v(z), \mu_{v(z)}). \tag{10.7}$$

where $\mu_y = \mu(\theta)$, for $\theta \in \mathcal{P}_y$. Then, we define a Q-function $Q : \Theta \times \mathcal{A} \to \mathbb{R}$ as

$$Q(\theta, a) := g^T(a)\theta + \beta \sum_{z \in \mathcal{Z}} V_\mu(T(\theta, z, a))\sigma(\theta, z, a).$$

According to Theorem 10.1, we can rewrite the Q-function as

$$Q(\theta, a) = \eta^T(\theta, a) \times \theta, \quad \text{with}$$

$$\eta(\theta, a) = g(a) + \beta \sum_{z \in \mathcal{Z}} \Lambda(a) \Phi(a) \eta(\theta', \mu).$$

where $\eta(\theta, \mu)$ is obtained by solving (10.7). We update the policy using

$$\mu(\theta) \in \arg \min_{\mu \in \mathcal{A}} Q(\theta, \mu). \tag{10.8}$$

Since the number of Eq. (10.7) is finite, we can use linear programming to solve (10.8).

Now, we aim to show the existence of an MP. To this end, we need the following preliminary definitions. We let $\mathcal{D}_\mu$ be the smallest closed set of discontinuities of stationary policy μ, i.e.,

$$\mathcal{D}_\mu := \{\theta \in \Theta : \mu(\theta) \text{ is discontinues at } \theta\}.$$

Given $D_0 := \mathcal{D}_\mu$, we define a sequence of sets $\mathcal{D}_1, \mathcal{D}_2, \ldots, \mathcal{D}_n, \ldots$, where $\mathcal{D}_n$ is given by

$$\mathcal{D}_n := \{\theta \in \Theta : T(\theta, z, \mu) \in \mathcal{D}_{n-1} \text{ for some } z \in \mathcal{Z}\}, \tag{10.9}$$

where $n \in \mathbb{Z}_+$ is the set index. Then, the following theorem presents a sufficient and necessary condition for the existence of an MP.

Theorem 10.2 (Sondik [156]) *Consider the discounted cost POMDP defined in (11.10):*

1. *A stationary policy is finitely transient if and only if there exists an integer n such that $\mathcal{D}_n = \emptyset$.*

2. *A finitely transient policy μ induces a Markov partition $\mathcal{P}$ of Θ. The elements in the set $\cup_{j=0}^n \mathcal{D}_j$ specify the boundaries of the partition $\mathcal{P}$.*

Remark 10.1 Theorem 10.2 provides a method to find an MP $\mathcal{P}$ of a POMDP, i.e., we need to show that there exists a finite constant n such that $\mathcal{D}_n = \emptyset$ for each stationary policy μ_τ, where $\tau = 0, 1, \ldots$ is the iteration step.

Hence, one way to solve an infinite horizon POMDP is to find an MP. In the next chapter, we will present a specific application based on a cyber POMDP. We will use Theorem 10.2 to show that the cyber POMDP admits an MP.

10.3 Conclusions and Notes

In this chapter, we have introduced the formulation of POMDPs and their structural properties. Discrete-state POMDPs are often used to capture the dynamics of the cyber state, while continuous state-space models are useful for physical layer control designs. In the next chapter, we will present a specific application based on POMDPs. In the application, the defender cannot observe the states, but it has a detector to diagnose the physical systems. Based on the application, we will present many analytical results, showing a strong interdependency between the cyber and physical layers. For a general theory of POMDPs, readers can refer to [82, 83, 131]. For a review of algorithmic solutions to POMDP, readers can refer to [130, 149].

One relevant and immediate extension of POMDPs is partially observable Markov games. The decision process is extended to a multiplayer scenario in which each player gathers his observations, forms his beliefs, and determines his control strategies. This class of problems is one type of decentralized stochastic control problem. Interested readers can refer to [36] for more recent advances in the area. One special class is a two-player one-sided partially observable zero-sum stochastic games. One player has the complete observation of the states while the other player has partial observations. Readers can refer to [60, 61] for recent developments on this problem.

Chapter 11
Secure and Resilient Control of ROSs

11.1 New Challenges in Networked ROSs

As the scale and scope of industrial robotics keep growing, directly controlling these robots becomes increasingly challenging due to their various structures and hardware. To meet these challenges, many robotic researchers have proposed several frameworks to organize different types of robots in an individual network, leading to numerous robotic software systems in industries [81]. A Robot Operating System (ROS) is one of the robotic software systems to provide operating system-like functionality on different computer clusters [132].

The future industrial operation and production will depend on open networks to facilitate data exchange between factories and data-backbend services, e.g., the outsourcing additive manufacturing [136]. This feature exposes the industrial facilities as well as the ROS to cyber threats [28]. A well-known example is the Stuxnet worm, a malicious computer worm that targets industrial control system [109]. These cyber attacks also threaten the industrial ROSs due to their sensitive information exchanged among various ROS nodes. Besides stealing confidential information, attackers can launch cyber attacks to damage the physical components of the ROS.

This work aims to analyze the impact of the false-data-injection attack on a ROS and its agents, such as multiple sensors, controllers, actuators, plants, and other devices. Figure 11.1 presents an application in which a ROS controls a car manufacturing process. The ROS master provides software frameworks (e.g., software tools and libraries) for operators and ROS agents to use [118]. For example, the operators and the sensors send their data to the ROS master through the ROS networks. To assemble the car, the robotic arms can receive data from the ROS networks by running a reading function provided by the ROS libraries. Due to the rapid development of the sophisticated attacks, it is challenging to use current

Q. Zhu, Z. Xu, *Cross-Layer Design for Secure and Resilient Cyber-Physical Systems*,
Advances in Information Security 81, https://doi.org/10.1007/978-3-030-60251-2_11

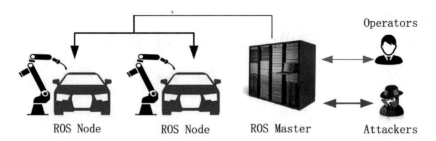

Fig. 11.1 An example of a ROS under cyber attacks: in the car-manufacturing line, operators send references to the ROS agents, while attackers inject a false data to damage the physical system

protections (e.g., isolated the ROS networks from open networks) to protect the ROS. Advanced attacks, e.g., Advanced Persistent Threats (APT), can exploit zero-day vulnerabilities of the system, stealthily steal passwords and cryptographic keys, compromise the specific target, and inject unauthorized data to the systems [39]. The false data can inflict severe damage on the physical components.

One conventional solution to assure data integrity and defend against a false-data injection is to use a Message Authentication Code (MAC). Given a message and a secret key, a sender can run the MAC algorithm to generate a tag such that the receiver can authenticate the data with the tag [73]. However, the primary challenge of implementing the conventional cryptography into ROSs is that many robotic systems are delay-sensitive, and the encryption scheme will introduce a significant delay, degrading the control performance. For instance, an encryption scheme RSA has a long key length and high processing requirements [115], but robotic arms might only endure a short delay (e.g., microseconds) [116]. Besides, a large-scale ROS might require multi-robot cooperation, in which the delay may aggregate significantly due to the dependence among a large number of robots.

To this end, for delay-sensitive systems, researchers have developed lightweight cryptography to reduce the computational complexity. For example, Yao et al. [181] have developed a lightweight attribute-based encryption for the Internet of Things (IoT) with limited computing resources. Similarly, we can also design a lightweight MAC (LMAC) for delay-sensitive ROSs by shortening the length of the secret key. However, the LMAC will increase the risk that the attacker compromises the security mechanism since the short-length key is accessible to break [186]. Besides designing the LMAC, we should also consider a resilient policy for the systems.

In this chapter, we design the key length of an LMAC for delay-sensitive ROSs to achieve data integrity. Given the risk of the LMAC, we define a cyber state to indicate whether the attacker compromises the LMAC. Since it is challenging to observe the cyber state, we develop a Chi-square detector to diagnose the LMAC. To find an optimal resilient defense strategy, recovering the LMAC after a successful attack, we formulate a cyber Partially Observed Markov Decision Process (POMDP) based on the proposed detector. Our analytical results characterize a threshold policy for the POMDP. Based on the POMDP, we also investigate the relationship between the detector's parameters and the cyber policies. Our numerical experiments indicate a strong interdependency between the physical parameters and the cyber strategies (e.g., the key length of the LMAC and the resilient defense strategy). For example, a more stable system (i.e., the eigenvalues of the discrete system matrix are close to zero) will choose a longer key length and a less conservative resilient strategy since it is less sensitive to the delay.

11.2 Problem Formulation

In this section, we first outline the steps of our design. Due to the delay introduced by the cryptography, we design a time-delay controller for the ROS agents. Then, we propose an attack model related to data integrity. Given the attack model, we develop a lightweight MAC (LMAC) and estimate the delay based on the key length. Finally, we define cyber states for the lightweight scheme and use a Markov Decision Process (MDP) to capture tradeoffs among security, resilience, and physical control performance of the ROS agents.

11.2.1 The Outline of the Proposed Mechanism

The proposed mechanism considers the tradeoffs among the physical performance, security, and resilience of the ROS agents. Figure 11.2 illustrates the architecture of the proposed mechanism. The solid blue arrow shows the forward interdependen-

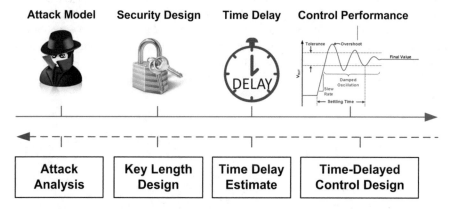

Fig. 11.2 The architecture of the proposed mechanism. The solid blue line shows the forward interdependencies: the operator uses a security mechanism to defend the attack, and the cryptography introduces a delay to the physical system, degrading the physical performance. The dashed red arrow shows the backward design: we design a delay controller, estimate the relationship between delay and key length, develop an LMAC based on physical requirements

cies. We consider a data-integrity attack, where the attacker aims to inject false data to a ROS agent. Given the attack model, the operator uses a security mechanism to protect the system, but the cryptography introduces a delay to the physical layer, degrading the control performance.

The dashed red arrow in Fig. 11.2 shows the steps of the design, i.e., we design the proposed mechanism from the physical layer to the cyber layer. Given a delay, we develop an optimal time-delay controller for the physical system. Based on the time-delay controller, we estimate the physical control cost. Then, we define a delay function to evaluate the relationship between the key length and delay. Finally, we formulate a cyber defense problem to capture the tradeoffs between control performance and security issues. Intuitively, a long-length key can achieve a high-level security defense, but it will introduce a significant delay to the physical layer. Hence, we need to design the proposed mechanism holistically, taking into account both the cyber and physical issues.

11.2.2   The Physical Dynamics of a ROS Agent

The first step of our design is to build a connection between the delay and the physical performance of a ROS. Hence, we consider a time-delay dynamical model:

$$x_{t+1} = Ax_t + Bu_{t-d} + w_t, \quad \forall x_0 \in \mathcal{X}, \tag{11.1}$$

where $t \in \mathbb{Z}_+$ is the physical time instant, and $d \in \mathbb{Z}_+$ is a given delay; $x_t \in \mathcal{X} \subset \mathbb{R}^{n_x}$ is the physical state with a given initial condition $x_0 \in \mathcal{X}$; $u_t \in \mathcal{U} \subset \mathbb{R}^{n_u}$ is the physical control input; $w_t \in \mathcal{X}$ is an additive Gaussian disturbance with a covariance matrix $\Sigma_w \in \mathbb{R}^{n_x \times n_x}$; $A \in \mathbb{R}^{n_x \times n_x}$ and $B \in \mathbb{R}^{n_x \times n_u}$ are constant matrices. Since all the signals are digital, $\mathcal{X}$ and $\mathcal{U}$ are quantized spaces. However, we do not consider the impacts of quantization errors in this work since the length of the transmitting message could be sufficiently long, e.g., 256 bits.

From $t = 0$ to $N - 1$, each ROS agent needs to track a given reference $r \in \mathcal{X}$. Hence, the control objective is to solve the following problem:

$$J_{\text{nor}}(x_0, r; d) := \min_{\mathbf{u}} \mathbb{E}\left\{ \sum_{j=0}^{d-1} \|A^j(x_0 - r)\|_F^2 + \sum_{t=d}^{N-1} \left(\|x_t - r\|_F^2 + \|u_{t-d}\|_R^2 \right) \right\},$$

subject to (11.1), where $\mathbf{u} = \{u_d, \ldots, u_N\}$ is the control sequence; $\|b\|_G^2 = b^T G b$ for $b \in \mathbb{R}^{n_b}$ and $G \in \mathbb{R}^{n_b \times n_b}$; $F \in \mathbb{R}^{n_x \times n_x}$ and $R \in \mathbb{R}^{n_u \times n_u}$ are positive symmetric matrices. The optimal delay control is given by

$$u_t^* = -(B^T P_t B + R)^{-1} B^T P_t A(\hat{x}_{t+d} - r),$$

where P_t is the solution to the algebraic Riccati equation, and $\hat{x}_{t+d}$ is an estimation, defined by

$$\hat{x}_{t+d} := A^d x_t + \sum_{j=t}^{t+d-1} A^{t+d-j-1} B u_{j-d}. \tag{11.2}$$

We can estimate the physical cost $J_{\text{nor}}(\cdot)$ in the normal case under a given delay. In the following part, we will define an attack model and estimate the physical cost in the attack case.

11.2.3 Attack Model: Data-Integrity Attack

For ROS-based applications, Bernhard et al. have proposed [41] three attack models related to data privacy, integrity, and availability. In this work, we focus on data integrity attacks since an unauthorized input can cause severe damage to robotic systems. To achieve this goal, an adversary first needs to compromise the security mechanism, e.g., using an APT-based attack to steal the cryptographic keys or exploit a zero-day vulnerability [39]. Following a successful cyber attack, the attacker can inject false data to mislead the ROS agent to track a fake reference $\bar{r} \in \mathcal{X}$.

Now, we formulate the attacker's problem, where the attacker aims to deviate the system from r to $\bar{r}$ by injecting an input $v_t \in \mathcal{U}$. Hence, the physical problem for the attacker is defined by

$$J_{\text{att}}(x_0, \bar{r}; d) := \min_{\mathbf{v}} \mathbb{E}\left\{ \sum_{j=0}^{d-1} \|A^j(x_0 - \bar{r})\|_F^2 \right.$$
$$\left. + \sum_{t=d}^{\infty} \left(\|x_t - \bar{r}\|_F^2 + \|v_{t-d} - u_{t-d}\|_R^2 \right) \right\}, \tag{11.3}$$

subject to (11.1), where $\mathbf{v} = [v_d, \ldots, v_N]$ is the injected data of the attacker. The third term $\|v_{t-d} - u_{t-d}\|_R^2$ in (11.3) indicates the stealthiness of the attacker. The solution to problem (11.3) is given by

$$v_t = u_t^* - (B^T P_t B + R)^{-1} B^T P_t A(\hat{x}_{t+d} - \bar{r}).$$

Here, we assume that the attacker can observe u_t^* during the attack. Substituting v_t^* into (11.3) also yields physical cost $J_{\text{att}}(x_0, \bar{r}; d)$ in the attack case. In the next subsection, we will employ an LMAC to mitigate the data-integrity attack.

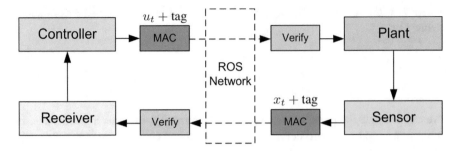

Fig. 11.3 The Message Authentication Code (MAC): the controller (sensor) uses the MAC to generate a tag before sending to the plant (receiver); the plant (receiver) uses the tag to verify the message

11.2.4 The Lightweight MAC and the Estimated Delay

To protect the ROS from the integrity attack, we design an LMAC to authenticate the data transmitting in the ROS networks. As shown in Fig. 11.3, senders (i.e., controllers or sensors) can use the LMAC algorithm to generate the tag, and the receiver (i.e., actuators or controllers) can verify the data based on the tag and message. It is essential that the LMAC gives a small overhead on the communications and computation, introducing a short delay to the ROS agents.

We use a classical approach to design an LMAC by shortening the key length. Suppose that the sender has a binary message $m \in [0, 1]^{\ell_m}$, where $\ell_m \in \mathbb{Z}_+$ is the message length. We define a MAC function $f_{\text{MAC}} : [0, 1]^{\ell_m} \times [0, 1]^{\ell_{key}} \rightarrow [0, 1]^{\ell_{tag}}$ to generate a tag, where $\ell_{key} \in \mathcal{L}_{key}$ is the key length, and $\ell_{tag} \in \mathbb{Z}_+$ is the tag length. Theorem 11.1 characterizes the risk level of an MAC.

Theorem 11.1 *Let $m, \bar{m} \in [0, 1]^{\ell_m}$ be two different messages. Given $\delta > 0$, we say that a MAC is δ-secure if*

$$\Pr(Ver_{sk}(m) = Ver_{sk}(\bar{m})) \leq \delta(\ell_{key}), \quad with \ \delta(\ell_{key}) := \ell_m \cdot (\sqrt{2})^{-\ell_{key}}, (11.4)$$

where $Ver_{sk}(\cdot)$ is the verification algorithm of the MAC, and sk is a secret key; $\delta : \mathcal{L}_{key} \rightarrow \mathbb{R}$ is a function to evaluate the risk level.

Remark 11.1 The proof of Theorem 11.1 can be founded in Chapter 4 of [73]. Theorem 11.1 states that an MAC fails to distinguish two different messages with a probability no greater than δ.

In general, message length ℓ_m (e.g., $\ell_m = 256$ bits) is fixed, so key length ℓ_{key} will affect risk level. To use (11.4) as our risk model, we need the following assumption:

Assumption 11.2 *Given a message length $\ell_m \in \mathbb{Z}_+$, we assume that key length ℓ_{key} satisfies that*

$$\ell_{key} > 2 \log_2(\ell_m).$$

Remark 11.2 Assumption 11.2 guarantees that $\delta(\cdot)$ is in the range $(0, 1)$. In fact, ℓ_{key} should be sufficiently long to ensure that the outcome of $\delta(\cdot)$ is negligible.

Now, we aim to estimate the delay based on a key length ℓ_{key}. The exact delay depends on many factors, such as time complexity of the MAC, the sampling time, the main frequency of the computing processor, etc. We illustrate an example of estimating the delay based on ℓ_{key}. Suppose that the central computing processor needs to run $\ell_m \times \ell_{key}^2$ times for the MAC. Then, we estimate d using

$$d = f_d(\ell_{key}) = \left\lceil \frac{\ell_m \cdot \ell_{key}^2}{freq_c \cdot T_s} \right\rceil, \quad \forall \ell_{key} \in \mathcal{L}_{key}, \tag{11.5}$$

where $d \in \mathbb{Z}_+$ is the total delay caused by the cryptography, and $f_d : \mathcal{L}_{key} \to \mathbb{Z}_+$ is the time-delay function; $\lceil \cdot \rceil$ is the ceiling function; $freq_c$ is the main frequency of the computing processor; T_s is the sampling time. In fact, we can obtain the exact values of the delay by running real experiments. In Sect. 11.5, we will use (11.5) to estimate the delay in the numerical experiments.

11.2.5 Physical-Aware Design of the Key Length

In this subsection, we aim to design a key length based on physical requirements. Since the key length is closely related to the delay, when designing a key length, we should take the physical needs into the highest priority.

For different applications, the ROS system will have different requirements for the delays. Here, we illustrate an example to show how to decide the time delay based on a given physical criterion. Suppose that we have a desirable control cost $J_{nor}^*(x_0, r; 0) > 0$. Then, we consider the following criteria:

$$J_{bias} := \frac{J_{nor}(x_0, r; d) - J_p^*(x_0, r; 0)}{J_p^*(x_0, r; 0)} \leq \bar{J}_p, \quad \forall x_0 \in \mathcal{X}, \tag{11.6}$$

where $d = f_d(\ell_{key})$, for $\ell_{key} \in \mathcal{L}_{key}$; $J_{bias} \geq 0$ is the bias of the physical cost; $\bar{J}_p$ is a threshold of the physical criterion. Hence, we can choose the longest key length ℓ_{key}^* in $\mathcal{L}_{key}$ to achieve the highest security while physical cost J_{nor} satisfying (11.6).

11.2.6 Cyber States and Cyber Actions

Since the lightweight MAC with a short key length will increase the risk that the attacker guesses the correct secret key, we define a cyber state $s \in \mathcal{S}$, for $\mathcal{S} = \{s_w, s_f\}$, to capture its status. State s_w stands for the working state, where the LMAC can protect the system, while state s_f stands for the failure state, where the LMAC fails to protect the system.

Figure 11.4 illustrates the transition of the cyber state. When $s = s_w$, we assume that the probability that the attacker succeeds in compromising the LMAC is ρ, named the risk level, and the probability that the system maintains secure is $1 - \rho$. When $s = s_f$, the probability that state s goes back to state s_w is γ, while the probability staying in state s_f is $1 - \gamma$.

In failure state s_f, we assume that the attacker has the secret key of the LMAC, i.e., the attacker can inject a fake data to pass the verification. To wipe out the impact of the attack, operators can reboot the cyber layer and update the security mechanism (e.g., update the secret keys of the LMAC) [2]. Although rebooting the cyber layer will induce a cost (e.g., the cyber system of ROS needs to wait for a specific period), the restoration of the cyber layer will ensure that the LMAC with a new key can protect the system from the future attack.

Based on the cyber state, we define two actions to formulate the cyber problem. The cyber action set is defined by $\mathcal{A} := \{a_s, a_r\}$. When choosing a_s, the agent will use the security mechanism (i.e., LMAC) to protect the system. When choosing a_r, the agent will reboot the cyber layer. We assume that the agent needs to pay a cyber cost $E_c > 0$ to reboot the cyber layer.

Given the above setups, we define transition functions $\rho(a)$ and $\gamma(a)$ as

$$\rho(a) := \begin{cases} \delta(\ell_{key}^*), & \text{if } a = a_s; \\ 0, & \text{if } a = a_r; \end{cases} \qquad \gamma(a) := \begin{cases} 0, & \text{if } a = a_s; \\ 1, & \text{if } a = a_r, \end{cases}$$

where $\delta(\cdot)$ is defined by (11.4), and ℓ_{key}^* is the optimal key length.

Given $\rho(a)$ and $\gamma(a)$, we define a transition matrix $\Lambda(a) \in \mathbb{R}^{2 \times 2}$ as

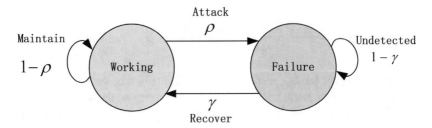

Fig. 11.4 The Cyber State of a ROS node: in the working state, the MAC can protect the ROS agent; in the failure state, the attacker compromises the MAC and attacks the system

$$\Lambda(a) := \left[\lambda(s'; s, a)\right]_{s', s \in \mathcal{S}, a \in \mathcal{A}} = \begin{bmatrix} 1 - \rho(a) & \rho(a) \\ \gamma(a) & 1 - \gamma(a) \end{bmatrix},$$

where $\lambda(s'; s, a) := \Pr[s'|s, a]$.

Given matrix $\Lambda(a)$, we will formulate a cyber MDP to capture the tradeoffs between the security and physical performance in the subsection. We aim to design an optimal cyber strategy to achieve the best performance holistically.

11.2.7 Stochastic Cyber Markov Decision Process

In this subsection, we will formulate a cyber problem to achieve an optimal resilient strategy. We first define a cyber time scale to distinguish the time scales between the physical and the cyber events. Secondly, we introduce an instantaneous cyber cost function. Based on the cyber cost function, we formulate a cyber MDP to find an optimal cyber policy for the ROS agent.

To define the cyber problem, we introduce a timescale k of the cyber events. The cyber activity timescale indicates the frequency of cyber defense actions (e.g., patching, rebooting, etc.). We assume that a ROS agent can accomplish a task within one step at the cyber time scale. This assumption also coincides with the concept of dwell time in hybrid systems [95]. Hence, the unit of k could be an hour or a day, since the frequency of patching and upgrade is often done daily or weekly for a server [133]. The unit of physical time scale t could be a microsecond. Figure 11.5 illustrates the relationship between the physical and cyber time scale. To simplify the problem, we assume that the reference r_k remains the same in each cyber time slot. For trajectory tracking problems, r_k should be a sequence of references.

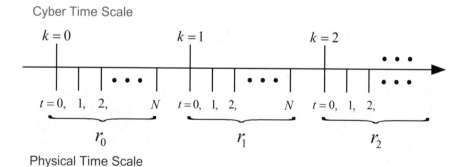

Fig. 11.5 The cyber and physical time scale: t represents the physical time scale, while k represents the cyber time scale

Given the cyber time scale, we define an instantaneous cyber cost function g : $\mathcal{S} \times \mathcal{A} \to \mathbb{R}$ to capture the tradeoffs between security and physical performance. The cost function is defined by

$$g(s, a) := \begin{cases} \alpha J_{\text{nor}} \cdot \mathbf{1}_{\{a=a_s\}} + E_c \cdot \mathbf{1}_{\{a=a_r\}}, & \text{if } s = s_w; \\ \alpha J_{\text{att}} \cdot \mathbf{1}_{\{a=a_s\}} + E_c \cdot \mathbf{1}_{\{a=a_r\}}, & \text{if } s = s_f, \end{cases}$$

where J_{nor} and J_{att} are short for $J_{\text{nor}}(x_0, r; d)$ and $J_{\text{att}}(x_0, \bar{r}; d)$, respectively; $\alpha > 0$ is a tuning coefficient; $E_c > 0$ is the cost for rebooting the cyber system. Note that physical costs J_{nor} and J_{att} are nonnegative, so $g(s, a)$ is also non-negative. Cost function $g(s, a)$ considers the tradeoffs between the physical performance and resilience of the system. For example, if J_{att} is significant, then the defender will reboot the cyber layer with a high frequency to mitigate the attack. However, if E_c is large, then the defender will be cautious to choose a_r since the rebooting cost is high.

One basic objective is to ensure that the ROS agent chooses a_r in state s_f. To this end, we make the following assumption on the cost functions.

Assumption 11.3 *Given $J_{nor}(\cdot)$ and $J_{att}(\cdot)$, we assume that $E_c > 0$, $\alpha > 0$, and $d \in \mathbb{Z}_+$ satisfy that*

$$\alpha J_{nor}(x_0, r; d) < E_c < \alpha J_{att}(x_0, \bar{r}; d),$$

for given $x_0, r, \bar{r} \in \mathcal{X}$.

Remark 11.3 Assumption 11.3 ensures that the ROS agent has the incentive to choose action a_s in state s_w and choose action a_r in state s_f.

Given function $g(\cdot)$, our objective is to find a cyber policy $\mu_{\text{MDP}} : \mathcal{S} \to \mathcal{A}$ to minimize an infinite horizon discounted problem

$$J_{\text{MDP}}(s_0, \mu_s) := \mathbb{E}_\pi \left\{ \sum_{k=0}^{\infty} \beta^k g(s_k, \mu_s) \middle| s_0 \right\}, \tag{11.7}$$

where $\beta \in (0, 1)$ is a discount factor.

Note that if the ROS can observe state s_k at each time k, we can quickly solve problem (11.7). However, in real applications, it is challenging to know if the attacker compromises the security mechanism, i.e., the ROS agent has incomplete information about cyber state s_k. To this end, we will formulate a Partially Observed Markov Decision Process (POMDP) to capture the uncertainty of state s_k. We will present details of the cyber POMDP in the following section.

11.3   Cyber POMDP Formulation for ROSs

In this section, we formulate a cyber POMDP based on a Chi-square detector. Then, we present theoretical results to find an optimal resilient defense strategy. We also analyze the relationship between the detection performance and the optimal defense strategy.

11.3.1   Basic Setups of the Cyber POMDP

Since it is challenging to observe state $s \in S$, we need to design a detector to diagnose the attack. In the attack model, the attacker aims to deviate the system from the original reference. One way to detect an abnormal output in dynamic systems with stochastic noises is by using a Chi-square detector, which can mitigate the impact of stochastic noise on the detection process [104].

In the physical control problem, we assume that N is sufficiently large. When t is closed to N, state x_t should be around reference r_k. Hence, we sample last N_s ($N_s < N$) number of state x_t. Then, we consider the mean value of squares:

$$\bar{\chi}_k^2 := \frac{1}{N_s} \sum_{t=0}^{N_s-1} (x_{N-t} - r_k)^T \Sigma_w^{-1} (x_{N-t} - r_k).$$

According to the Wald test (See Sections 9.2, 10.5 in [55]), $\bar{\chi}^2$ satisfies a Chi-square distribution with n_x (the dimension of x_t) degrees.

Without attacks, $\bar{\chi}_k^2$ should be small. Hence, we define an observation $z \in \mathcal{Z}$, for $\mathcal{Z} := \{z_q, z_u\}$, where z_q means the physical performance satisfies the requirement, and z_u means otherwise. We obtain the z_k using

$$z_k = \begin{cases} z_q, & \text{if } \bar{\chi}_k^2 \leq \bar{z}; \\ z_u, & \text{if } \bar{\chi}_k^2 > \bar{z}, \end{cases} \tag{11.8}$$

where $\bar{z} > 0$ is a detection threshold. Given detector (11.8), we define a detection matrix, given by

$$\Phi(z) := \begin{bmatrix} \phi_w(z) & 0 \\ 0 & \phi_f(z) \end{bmatrix}, \quad \text{for } z \in \mathcal{Z},$$

where $\phi_w(z) := \Pr(z|s = s_w)$ and $\phi_f(z) := \Pr(z|s = s_f)$. We can compute the above elements in $\Phi(z)$ using

$$\phi_w(z_q) = \Pr(\bar{\chi}^2 \leq \bar{z}|s = s_w), \quad \phi_w(z_u) = 1 - \phi_w(z_q),$$

$$\phi_f(z_u) = \Pr(\bar{\chi}^2 > \bar{z}|s = s_f), \quad \phi_f(z_q) = 1 - \phi_f(z_u).$$

158 11 Secure and Resilient Control of ROSs

The specific values of $\Pr(\bar{\chi}^2 \leq \bar{z}|s = s_w)$ and $\Pr(\bar{\chi}^2 > \bar{z}|s = s_f)$ depend on threshold $\bar{z}$, difference $\|r_k - \bar{r}_k\|_2$, and physical size n_x. In the numerical experiment, we will present an example to compute these parameters. According to detection theory, $\phi_w(z_q)$ is the True Negative (TN) rate since z_q means no attack while $s = s_w$. $\phi_f(z_q)$ is the Missed Detection (MD) rate since z_q means no attack while $s = s_f$. In the numerical experiments, we can plot the Receiver Operating Characteristic (ROC) curves based on different values of $\bar{z}$, r_k, and $\bar{r}_k$.

Given detector (11.8), we establish a belief of cyber state s_k. Let $\theta_0 := \Pr(s_0 = s_w)$ be the initial probability of staying in state s_w. At cyber time k, the ROS agent has an information set

$$i_k := \{\theta_0, a_0, z_1, \ldots, a_{k-1}, z_k\}. \tag{11.9}$$

Then, we define a posterior belief $\theta_k := \Pr(s_k = s_w|i_k)$. We call θ_k as the belief state, for $\theta_k \in \Theta := [0, 1]$. To update θ, we use Bayes' rule to develop a belief update function, i.e., $\theta_{k+1} = T(z_{k+1}, \theta_k, a_k)$, where $T : \mathcal{Z} \times \Theta \times \mathcal{A} \to \Theta$ is defined by

$$T(z, \theta, a) := \frac{[1, 0] \times \Phi(z)\Lambda^T(a)[\theta, 1 - \theta]^T}{\sigma(z, \theta, a)}, \text{ with}$$

$$\sigma(z, \theta, \mu) := [1, 1] \times \Phi(z)\Lambda^T(a)[\theta, 1 - \theta]^T.$$

More details about general POMDP can be found in [82]. Similar to (10.4), our objective is to find a cyber policy $\mu : \Theta \to \mathcal{A}$ to minimize the expected cost function

$$J_c(\theta_0, \mu) = \mathbb{E}_\mu\left\{\sum_{k=0}^{\infty} \beta^k \bar{g}(\theta_k, \mu)\bigg|\theta_0\right\}, \tag{11.10}$$

where $\bar{g}(\theta, a) := \theta g(s_w, a) + (1 - \theta)g(s_f, a)$. Then, we define a value function $V_\mu : \Theta \to \mathbb{R}$:

$$V_\mu(\theta_k) := \mathbb{E}_\mu\left\{\sum_{j=k}^{\infty} \beta^{j-k} \bar{g}(\theta_j, \mu)\bigg|\theta_k\right\}$$

$$= \bar{g}(\theta_k, \mu) + \beta \sum_{z \in \mathcal{Z}} V_\mu(T(\theta_k, z, \mu))\sigma(\theta_k, z, \mu),$$

where the second equality of comes from the Bellman's dynamic programming equation. However, since belief space Θ is uncountable, the above dynamic programming is challenging to solve. Hence, in the next subsection, we present our main results that characterize an optimal threshold policy for (11.10).

11.3.2 Main Results of Cyber POMDP

In this part, we will study the properties of cyber POMDP (11.10) and propose an algorithmic solution to solve the problem. The following theorem presents the first property of the POMDP.

Theorem 11.2 *Consider the POMDP with a finite $\mathcal{S}$, a finite $\mathcal{A}$, and a finite $\mathcal{Z}$. Given a stationary policy μ, $V_\mu(\theta)$ is piecewise linear in $\theta \in \Theta$, i.e.,*

$$V_\mu(\theta) = \theta\eta(s_w, \theta, \mu) + (1 - \theta)\eta(s_f, \theta, \mu), \tag{11.11}$$

and $\eta_\mu : \mathcal{S} \times \Theta \to \mathbb{R}$ satisfies the linear equations:

$$\begin{bmatrix} \eta(s_w, \theta, \mu) \\ \eta(s_f, \theta, \mu) \end{bmatrix} = \begin{bmatrix} g(s_w, \mu) \\ g(s_f, \mu) \end{bmatrix} + \beta \sum_{z \in \mathcal{Z}} \Lambda_\mu \Phi_z \begin{bmatrix} \eta(s_w, \theta', \mu) \\ \eta(s_f, \theta', \mu) \end{bmatrix}, \tag{11.12}$$

where $\theta' := T(z, \theta, \mu)$.

Proof We use iteration to prove Theorem 11.2. Given a policy μ, we start with $V_{\mu,0}(\theta) = 0$. We observe that

$$V_{\mu,1} := \theta g(s_w, \mu) + (1 - \theta)g(s_f, \mu).$$

Hence, $V_{\mu,1}$ is piecewise linear in θ. We assume that

$$V_{\mu,\tau-1} = \theta\eta_{\tau-1}(s_w, \theta) + (1 - \theta)\eta_{\tau-1}(s_f, \theta).$$

Then, we can obtain

$$V_{\mu,\tau} = \theta\eta_\tau(s_w, \theta) + (1 - \theta)\eta_\tau(s_f, \theta), \quad \text{with}$$

$$\begin{bmatrix} \eta_\tau(s_w, \theta, \mu) \\ \eta_\tau(s_f, \theta, \mu) \end{bmatrix} = \begin{bmatrix} g(s_w, \mu) \\ g(s_f, \mu) \end{bmatrix} + \beta \sum_{z \in \mathcal{Z}} \Lambda_\mu \Phi_z \begin{bmatrix} \eta_{\tau-1}(s_w, \theta', \mu) \\ \eta_{\tau-1}(s_f, \theta', \mu) \end{bmatrix}.$$

Since $\beta \in (0, 1)$, the iteration will converge.   □

Although Theorem 11.2 presents a piecewise linear property of the POMDP, it is still challenging to use dynamic programming recursion to solve the problem due to the continuum property of $\theta \in \Theta$. The following definition facilitates the simplification of the POMDP. Given an MP, value function $V_\mu(\theta)$ becomes

$$V_\mu(\theta) = \theta\eta(s_w, y, \mu_y) + (1 - \theta)\eta(s_f, y, \mu_y),$$

where $\mu_y := \mu(\theta)$, for $\theta \in \mathcal{P}_y$, and $\eta(s, y, \mu_y)$ satisfies the following $2Y$ linear equations:

$$\begin{bmatrix} \eta(s_w, y, \mu_y) \\ \eta(s_f, y, \mu_y) \end{bmatrix} = \begin{bmatrix} g(s_w, \mu_y) \\ g(s_f, \mu_y) \end{bmatrix} + \beta \sum_{z \in \mathcal{Z}} \Lambda(\mu_y) \Phi_z \begin{bmatrix} \eta(s_w, v(z), \mu_{v(z)}) \\ \eta(s_f, v(z), \mu_{v(z)}) \end{bmatrix} \quad (11.13)$$

If our POMDP admits an MP, we can first solve equations (11.13) at iteration step $\tau - 1$. At iteration step τ, we can find the optimal policy using

$$\mu_\tau^*(\theta) \in \arg\min_{a \in \mathcal{A}} Q_\tau(\theta, a), \quad (11.14)$$

where $Q_\tau : \Theta \times \mathcal{A} \to \mathbb{R}$ is a Q-function, defined by

$$Q_\tau(\theta, a) := \bar{g}(\theta, a) + \beta \sum_{z \in \mathcal{Z}} V_{\mu_{\tau-1}}(\theta') \sigma(\theta, z, a). \quad (11.15)$$

Before showing the existence of an MP, we present two theorems to characterize the important properties of our problem. The first one shows the monotonicity of functions $V_\mu(\theta)$ and $Q(\theta, a)$. The second one shows that stationary policy μ_τ has a threshold structure at each step τ.

Theorem 11.3 *Given a stationary policy* $\mu : \Theta \to \mathcal{A}$, *value function* $V_\mu(\theta)$ *is non-increasing in* θ, *and Q-function* $Q(\theta, a)$ *is also non-increasing when* $a \in \mathcal{A}$ *is fixed. Function* $\eta(s, \theta, a)$ *satisfies that*

$$\eta(s_w, \theta, a_s) \le \eta(s_w, \theta, a_r) = \eta(s_f, \theta, a_r) \le \eta(s_f, \theta, a_s). \quad (11.16)$$

Proof Given Theorem 11.2, we can compute $\eta_\tau(s, \theta, \mu)$ iteratively, with $\eta_0(s, \theta, \mu) = g(s, \mu)$. Based on Assumption 11.3, we can see that

$$g(s_f, a) \ge g(s_w, a), \quad \forall a \in \mathcal{A},$$

where equality holds when $a = a_r$. We can verify that $\eta_0(s, \theta, \mu)$ satisfies (11.16). Assume that $\eta_{\tau-1}(s, \theta, \mu)$ satisfies (11.16). Using (11.12), we can show that $\eta_\tau(s, \theta, \mu)$ satisfies (11.16), and the iteration will converge due to $\beta \in (0, 1)$.

Given (11.16), it clear that $V_\mu(\theta)$ is non-increasing in θ. Similarly, we can show that $Q(\theta, a)$, given by

$$Q(\theta, a) = \theta \eta(s_w, \theta, a) + (1 - \theta) \eta(s_f, \theta, a),$$

is also non-increasing in θ when $a \in \mathcal{A}$ is fixed.   □

Theorem 11.3 implies a threshold structure of the optimal policy μ^*. The following theorem characterizes a structural result of policy μ_τ at each iteration step τ.

Theorem 11.4 *For each iteration step* τ, *the optimal stationary policy* μ_τ *of the POMDP (11.10) has a threshold structure, given by*

$$\mu_\tau(\theta) = \begin{cases} a_r, & \textit{if } \theta \leq \xi_\tau; \\ a_s, & \textit{if } \theta > \xi_\tau; \end{cases} \tag{11.17}$$

where $\xi_\tau \in [0, 1]$ is a threshold at the τ-th iteration.

Proof We first let $\theta = 0$ and compute

$$Q_\tau(0, a)$$
$$= g(s_f, a) + \beta \eta(s_w, 0, a)\gamma(a) + \beta \eta_\tau(s_f, 0, a)(1 - \gamma(a)). \tag{11.18}$$

According to Assumption 11.2 and Theorem 11.3, we can verify that $Q_\tau(0, a_s) \geq Q_\tau(0, a_r)$. Similarly, we can use the same steps to show that $Q_\tau(1, a_s) \leq Q_\tau(1, a_r)$. Since $Q(\theta, a)$ is continuous and monotonic in θ, there must exist a unique point $\xi_\tau \in [0, 1)$ such that $Q_\tau(\xi_\tau, a_s) = Q_\tau(\xi_\tau, a_r)$ Hence, policy μ_τ must have structure result (11.17) with a threshold ξ_τ, which completes the proof. $\square$

Theorem 11.4 illustrates that the policy $\mu(\theta)$ is a threshold policy with only one discontinuous point, i.e., $d_\mu = \{\xi_\tau\}$. Based on the above results, the following theorem shows that the cyber POMDP admits an MP.

Theorem 11.5 *The cyber POMDP defined by (11.10) admits a Markov Partition (MP) with a stationary policy μ_τ defined by (11.17) at each iteration τ.*

Proof To prove Theorem 11.5, we need to show that $D_n = \emptyset$ for a finite number $n \in \mathbb{Z}_+$. Given Theorem 11.4, we have $\mathcal{D}_0 = \{\xi_\tau\}$.

According to the definition of D_1, we need to find a belief $\theta_1 \in \Theta$ such that $T(z, \theta, \mu_\tau) = \xi_\tau$. Suppose that $\theta \leq \xi_\tau$. According to policy (11.17), we have $\mu_\tau(\theta) = a_r$. Hence, function $T(\cdot)$ becomes

$$T(z, \theta, a_r) = \frac{\phi_w \theta + \phi_w(1 - \theta)}{(\phi_w - \phi_f) + \phi_f} = 1 > \xi_\tau,$$

where ϕ_w and ϕ_f are short for $\phi_w(z)$ and $\phi_f(z)$, respectively. Hence, we conclude that no θ_1, for $\theta \leq \xi_\tau$, satisfies $T(z, \theta, a) = \xi_\tau$.

Then, we assume that there exists a belief $\theta > \xi_\tau$ such that $T(\theta, z, \mu_\tau) = \xi_\tau$. Given policy (11.17), we have $\mu_\tau(\theta) = a_s$, and

$$T(z, \theta, a_s) = \frac{\phi_w(1 - \rho_s)\theta}{(\phi_w - \phi_f)(1 - \rho_s)\theta + \phi_f},$$

where ρ_s is short for $\rho(a_s)$. Then, we observe that

$$\frac{\partial}{\partial\theta} T(\theta, \mu_\tau, z) = \frac{\phi_w \phi_f(1 - \rho_s)}{[(\phi_w - \phi_f)(1 - \rho_s)\theta + \phi_f]^2} > 0,$$

where the inequality comes from the facts that $\rho_s, \phi_w, \phi_f \in (0, 1)$. Hence, the function $T(z, \theta, \mu_\tau)$ is strictly increasing in θ when $\theta > \xi_\tau$.

Suppose that there exists $\theta_1, \theta_2 > \xi_\tau$ such that

$$\xi_\tau = T(\theta_1, \mu_\tau, z), \text{ and } \theta_1 = T(\theta_2, \mu_\tau, z).$$

Since $T(\theta, \mu_\tau, z)$ is strictly increasing in θ, for $\theta > \xi_\tau$, we arrive at $\theta_2 > \theta_1 > \xi_\tau$. If we keep finding θ_n with $\theta_{n-1} = T(\theta_n, \mu_\tau, z)$, we obtain that

$$\theta_{n-1} = T(\theta_n, \mu_\tau, z) \Rightarrow \theta_n > \theta_{n-1}. \tag{11.19}$$

Therefore, there must exist an n_τ^* such that $\theta_{n_\tau^*-1} = T(\theta_{n_\tau^*}, \mu_\tau, z)$ and $\theta_{n_\tau^*} > 1$, i.e., $d_{n_\tau^*} = \emptyset$ since $\theta_{n_\tau^*} \notin \Theta$. According to Theorem 10.2, the cyber POMDP admits an MP, which completes the proof. $\square$

Remark 11.4 Theorem 11.5 shows the existence of an MP in our problem. We can propose an algorithm, similar to the one in [156], to search the optimal policy μ^*.

Given an MP, we only need to solve $2Y$ linear equations given by (11.13) and find the optimal policy by solving (11.14) at each iteration step. In the following subsection, we investigate the relationship between the detection performance and the resilient strategy. We propose a special case based on a desirable detector to simplify the computation of the POMDP.

11.3.3   Special Case of the Cyber POMDP

In this subsection, we will analyze the interdependencies between the detector's parameters (i.e., the elements in $\Phi(z)$) and the properties of POMDP (11.10). We propose criteria to evaluate a desirable detector for the ROS agent. We will show that the desirable detector can further simplify the cyber POMDP problem.

Before analyzing the detector, we discuss a special case of the POMDP with a two-interval MP, i.e., $\mathcal{P} = \{\mathcal{P}_0, \mathcal{P}_1\}$. The following theorem characterizes a sufficient condition for the existence of a two-interval MP.

Theorem 11.6 *In the proposed cyber POMDP, the Markov Partition (MP) has two intervals, i.e., $\mathcal{P} = \{\mathcal{P}_0, \mathcal{P}_1\}$, where $\mathcal{P}_0 := \{\theta : 0 \le \theta < \xi\}$, $\mathcal{P}_1 := \{\theta : \xi \le \theta \le 1\}$, and $\xi \in (0, 1)$ is the threshold, if $T(\cdot)$ satisfies*

$$\forall \theta \in \mathcal{P}_1, \ T(z_q, \theta, \mu) \in \mathcal{P}_1, \tag{11.20}$$

$$\forall \theta \in \mathcal{P}_1, \ T(z_u, \theta, \mu) \in \mathcal{P}_0, \tag{11.21}$$

$$\forall \theta \in \mathcal{P}_0, z \in \mathcal{Z}, \ T(z, \theta, \mu) \in \mathcal{P}_1, \tag{11.22}$$

where μ is the threshold policy defined by (11.17).

Proof Given (11.17) and partition $\{\mathcal{P}_0, \mathcal{P}_1\}$, we have

$$\mu(\theta) = \begin{cases} a_r & \text{if } ,\theta \in \mathcal{P}_0; \\ a_s & \text{if } ,\theta \in \mathcal{P}_1. \end{cases}$$

Then, we can verify that $\{\mathcal{P}_0, \mathcal{P}_1\}$ satisfies the MP's properties stated in Definition 10.1 if (11.20)–(11.22) hold. $\qquad\square$

Based on the results of Theorem 11.6, we use the following theorem to build a connection between the detector's parameters (e.g., ϕ_w and ϕ_f) and the two-interval MP.

Theorem 11.7 *Suppose that we have a threshold policy μ defined by (11.17). The cyber POMDP has a two-interval MP specified by (11.20)–(11.22), if the elements of matrix $\Phi(z)$ satisfy the following conditions:*

$$\frac{\phi_w(z_q)}{\phi_f(z_q)} \geq \frac{1 + (\rho_s - 1)\xi}{(1 - \rho_s)(1 - \xi)}, \quad \frac{\phi_w(z_u)}{\phi_f(z_u)} < \frac{\xi\rho_s}{(1 - \rho_s)(1 - \xi)}, \quad (11.23)$$

where ρ_s is short for $\rho(a_s)$.

Proof Note that $\mu(\theta) = a_r$ if $\theta \in \mathcal{P}_0$. Then, we observe that

$$T(\theta, z, a_r) = \frac{\phi_w\theta + \phi_w(1 - \theta)}{(\phi_w - \phi_f) + \phi_f} = 1 \in \mathcal{P}_1.$$

Hence, we can see that (11.22) holds. Given $\mu(\theta) = a_s$, for $\theta \in \mathcal{P}_1$, we observe that

$$\begin{aligned} T(\theta, z_q, a_s) &= \frac{[1, 0] \times \Phi(z_q)\Lambda^T(a_s) \times [\theta, 1 - \theta]^T}{\mathbf{1}_2^T \Phi(z_q)\Lambda^T(a_s)[\theta, 1 - \theta]^T} \\ &= \frac{\phi_w(z_q)(1 - \rho_s)\theta}{\phi_w(z_q)(1 - \rho_s)\theta + \phi_f(z_q)(\rho_s\theta + 1 - \theta)}. \end{aligned}$$

Note that (11.20) is equivalent to that $T(\theta, z_q, a_s) \geq \xi$, for any $\theta \in \mathcal{P}_1$. Then, we arrive at

$$\frac{\phi_w(z_q)}{\phi_f(z_q)} \geq \max_{\theta \in \mathcal{P}_1} \left\{ \frac{\xi(\rho_s - 1)\theta + \xi}{(1 - \rho_s)(1 - \xi)\theta} \right\} = \frac{\xi(\rho_s - 1) + 1}{(1 - \rho_s)(1 - \xi)}.$$

Similarly, (11.21) is equivalent to that $T(\theta, z_u, a_s) < \xi$, for any $\theta \in \mathcal{P}_1$, i.e.,

$$\frac{\phi_w(z_u)}{\phi_f(z_u)} < \min_{\theta \in \mathcal{P}_1} \left\{ \frac{\xi(\rho_s - 1)\theta + \xi}{(1 - \rho_s)(1 - \xi)\theta} \right\} = \frac{\xi\rho_s}{(1 - \rho_s)(1 - \xi)}.$$

Given Theorem 11.6, the cyber POMDP admits the two-interval MP. $\qquad\square$

Remark 11.5 Theorem 11.7 presents conditions for the existence of a two-interval MP. These conditions depend on detection matrix $\Phi(z)$, security ratio $\rho(a_s)$, and threshold ξ. If we fix $\rho(a_s)$ and ξ, we can view (11.23) as two basic requirements of a desirable detector.

Given Theorem 11.7, we present a corollary to show another condition for the existence of the two-interval MP.

Corollary 11.1 *The cyber POMDP has a two-interval MP, if*

$$\xi \in \left[\frac{(1 - \rho_s)\phi_{w,u}}{\rho_s \phi_{f,u} + (1 - \rho_s)\phi_{w,u}}, \frac{(1 - \rho_s)\phi_{w,q} - \phi_{f,q}}{(1 - \rho_s)(\phi_{w,q} - \phi_{f,q})} \right], \tag{11.24}$$

where $\phi_{w,q}$, $\phi_{w,u}$, $\phi_{f,q}$, and $\phi_{f,u}$ are short for $\phi_w(z_q)$, $\phi_w(z_u)$, $\phi_f(z_q)$, and $\phi_f(z_u)$, respectively.

Proof The proof of Corollary 11.1 immediately comes from Theorem 11.7.   □

Corollary 11.1 provides another way to verify the existence of the two-interval MP. Based on the results Theorems 11.6 and 11.7, we use the following theorem to characterize a closed-form solution of threshold ξ if the POMDP admits the two-interval MP.

Theorem 11.8 *If the cyber POMDP has the two-interval MP satisfying (11.20)–(11.22), then we have the following closed-form solutions:*

$$\begin{bmatrix} \eta(s_w, 0, a_r) \\ \eta(s_f, 0, a_r) \\ \eta(s_w, 1, a_s) \\ \eta(s_f, 1, a_s) \end{bmatrix} = M^{-1} \begin{bmatrix} g(s_w, a_r) \\ g(s_f, a_r) \\ g(s_w, a_s) \\ g(s_f, a_s) \end{bmatrix}, \tag{11.25}$$

$$\xi = \frac{\eta(s_f, 1, a_s) - \eta(s_f, 0, a_r)}{\eta(s_f, 1, a_s) - \eta(s_f, 0, a_r) + \eta(s_f, 0, a_r) - \eta(s_f, 1, a_s)}, \tag{11.26}$$

$$M := \begin{bmatrix} I_2 - \beta \Lambda(a_r)\Phi(z_u) & -\beta \Lambda(a_r)\Phi(z_q) \\ -\beta \Lambda(a_s)\Phi(z_u) & I_2 - \beta \Lambda(a_s)\Phi(z_q) \end{bmatrix}. \tag{11.27}$$

Proof If the cyber POMDP has the two-interval MP satisfying (11.20)–(11.22), then we can rewrite (11.13) as

$$\begin{bmatrix} \eta(s_w, 0, a_r) \\ \eta(s_f, 0, a_r) \end{bmatrix} = \begin{bmatrix} g(s_w, a_r) \\ g(s_f, a_r) \end{bmatrix} + \beta \Lambda(a_r)$$

$$\times \left(\Phi(z_u) \begin{bmatrix} \eta(s_w, 0, a_r) \\ \eta(s_f, 0, a_r) \end{bmatrix} + \Phi(z_q) \begin{bmatrix} \eta(s_w, 1, a_s) \\ \eta(s_f, 1, a_s) \end{bmatrix} \right),$$

$$\begin{bmatrix} \eta(s_w, 1, a_s) \\ \eta(s_f, 1, a_s) \end{bmatrix} = \begin{bmatrix} g(s_w, a_s) \\ g(s_f, a_s) \end{bmatrix} + \beta \Lambda(a_s)$$

$$\times \left(\varPhi(z_u) \begin{bmatrix} \eta(s_w, 0, a_r) \\ \eta(s_f, 0, a_r) \end{bmatrix} + \varPhi(z_q) \begin{bmatrix} \eta(s_w, 1, a_s) \\ \eta(s_f, 1, a_s) \end{bmatrix} \right).$$

We can rewrite the above four equations into the following matrix form and obtain closed-form solution (11.25). Now, we present the closed-form solution of the threshold $\xi \in (0, 1)$. When $\theta = \xi$, the Q-function is given by

$$Q(\xi, a_r) := \xi\eta(s_w, 0, a_r) + (1 - \xi)\eta(s_f, 0, a_r)$$

$$Q(\xi, a_s) := \xi\eta(s_w, 1, a_s) + (1 - \xi)\eta(s_f, 1, a_s)$$

Equality $Q(\xi, a_r) = Q(\xi, a_s)$ yields the closed-form solution of ξ, given by (11.26). According to Theorem 11.4, ξ, defined by (11.26), stays in the range (0, 1). □

Remark 11.6 Theorem 11.8 presents a closed-form solution for the threshold policy. In real applications, we can first compute (11.25) and (11.26) without using iteration. If threshold ξ satisfies (11.24), then we can use ξ as the threshold.

Given the detector satisfying (11.23), the operator can solve the POMDP without using iteration. In real applications, it is simple to implement the proposed mechanism since both the physical and cyber policies are stationary. The ROS agent only needs to run the LMAC, compute the Chi-square values, and update belief θ. Designers can first estimate the delay by running experiments. After that, they can determine suitable key lengths based on specific requirements; i.e., our mechanism is flexible for various delay-sensitive ROS agents.

11.4 Experimental Results

In this section, we use a DC motor as a ROS agent to study and analyze the performance of the proposed mechanism. The experiments have two parts. In the first part, we focus on the physical layer. We investigate the tradeoff between security and control performance. Given the delay requirement, we compute the optimal key length for the system. In the second part, we use the proposed algorithm to derive the resilient policy, i.e., the threshold policy of the cyber layer. We analyze how the accuracies (e.g., the missed detection rate and true negative rate) of the detector impact the cyber performance, i.e., whether the system can reboot timely when the attacker compromises the security mechanism. Besides, we analyze how the physical parameters affect the cyber policy, showing a strong interdependency between the cyber and physical layers.

Table 11.1 Parameters of
the DC Motor

Parameters	Description	Values
R_a	Armature resistance	1.75 Ω
L_a	Armature inductance	2.83×10^{-3} H
I_r	Rotor inertia	0.03 kg m^2
K_v	Velocity constant	0.093 Vsec/rad
K_t	Torque constant	0.092 Nm/A
b_d	Damping coefficient	5.0×10^{-3} Nm s

11.4.1   Part I: Physical Performance

In this part, we test the tracking performance of a DC motor under our security
mechanism. The dynamic model of a permanent magnet DC motor is given by the
following two differential equations:

$$\frac{d}{dt}i_a = -\frac{R_a}{L_a}i_a - \frac{K_v}{L_a}\omega_a + \frac{V_a}{L_a}, \qquad \frac{d}{dt}\omega_a = \frac{K_t}{I_r}i_a - \frac{b_d}{I_r}\omega_a - \frac{T_L}{I_r},$$

where ω_a is the rotational velocity, and i_a is the armature current; V_a is the voltage,
and T_L is the torque of the load. Other parameters are constant, and their descriptions
and values are presented in Table 11.1.

Since velocity ω_a and current i_a are the variables, we define that $x := [\omega_a, i_a]^T$,
$u := [T_L, V_a]^T$. Then, we have the following linear continuous-time model:

$$\dot{x} = A_c x + B_c u, \tag{11.28}$$

where

$$A_c = \begin{bmatrix} -\frac{R_a}{L_a} & -\frac{K_v}{L_a} \\ \frac{K_t}{I_r} & -\frac{b_d}{I_r} \end{bmatrix}, \quad B_c = \begin{bmatrix} 1/L_a & 0 \\ 0 & -1/I_r \end{bmatrix}.$$

Given $T_s = 10^{-5}$ s, we discretize continuous-time model (11.28) to derive
model (11.1). Then, we study how the delay will affect the physical tracking performance. We set up two references at different time interval for the motor rotational
velocity. We assume that we have a key length space $\mathcal{L}_{key} = \{0, 10, 15, 20\}$.
Suppose that $freq_c = 3.2 \times 10^8$ Hz and $\ell_m = 256$ bits. Using (11.5) we obtain the
delays, given by $d \in \{0, 8, 18, 32\}$.

Based on the above setup and the given references (the tracking speed of
the motor), we simulate with two distinct disturbances, which are different in
their covariance matrix Σ_w. Figures 11.6 and 11.7 illustrate the results of the
tracking performance with different Σ_w. In both figures, we can see that the longer
the delay is, the worse the tracking performance will be. Besides, the variance
of the disturbances in Fig. 11.7 is larger than that in Fig. 11.6. By comparing
with the tracking performances in these two figures, we can see that the system

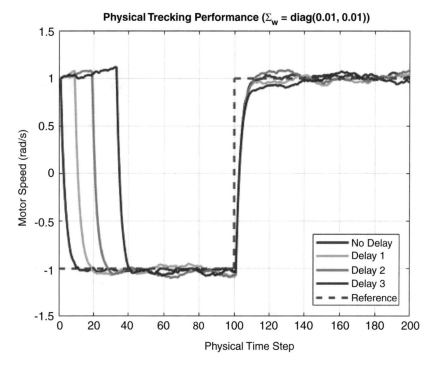

Fig. 11.6 The tracking performance of the DC motor with different delays under a small disturbance

output oscillates more sharply in Fig. 11.7 than the one in Fig. 11.6, i.e., the disturbance with a large variance, can strengthen the impact of the delay on the physical performance. The reason is that a large-variance disturbance deteriorates the estimation value $\hat{x}_t$ defined by (11.2), and the poor estimation of the state degrades the control performance.

11.4.2 Part II: Cyber Performance

In the second part of the experiments, we aim to evaluate the cyber performance under the proposed mechanism. Before doing that, we also care about how the physical parameters will affect the cyber policies since the cyber and physical layers are highly dependent. To this end, we change the settings in matrix A_c, leading to different eigenvalues of A in model (11.1), presented in Table 11.2. We can see that except Agent I, the systems of other Agents are unstable. Besides, one of the eigenvalues increases from Agent 1 to 5, while the other one is 1.

Figure 11.8a is a semilog plot, describing the interdependencies between physical costs and key length with different system parameters. We can see that the physical

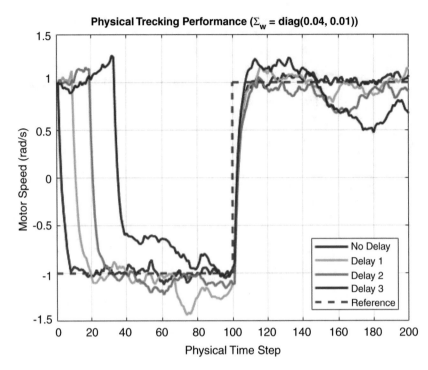

Fig. 11.7 The tracking performance of the DC motor with different delays under a large disturbance

Table 11.2 Agents' models with different eigenvalues

Agent index	Eigenvalues of matrix A
Agent 1	(0.9938, 1.0000)
Agent 2	(1.0178, 1.0000)
Agent 3	(1.0360, 1.0000)
Agent 4	(1.0507, 1.0000)
Agent 5	(1.0657, 1.0000)

costs grow exponentially with the key length, i.e., the key length has a significant impact on the physical costs. Besides, the physical parameters affect the increment of the costs. When the magnitude of the eigenvalues grows, physical cost J_{nor} grows exponentially, i.e., unstable systems are more sensitive to the delay. Figure 11.8b presents the relationship between the risk level ρ and key length. A long key length leads to a small risk level ρ, protecting the agent from attacks.

Given the physical tests, we will find the optimal key length for different agents. Suppose that the physical threshold is $\bar{J}_p = 7.8$ in the experiments. Figure 11.9 illustrates the values of J_{bias} with different key lengths, and the red dashed line is the value of $\bar{J}_p$. According to results in Fig. 11.9 and physical criteria (11.6), we obtain the optimal key lengths, i.e., $\ell_{key}(1) = 29$, $\ell_{key}(2) = 24$, $\ell_{key}(3) = 21$, $\ell_{key}(4) = 20$, $\ell_{key}(5) = 18$, where $\ell_{key}(i)$ is the optimal key length for agent i.

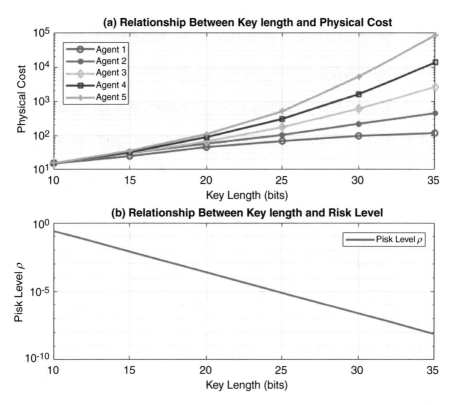

Fig. 11.8 (**a**) The relationships between key length and physical costs; (**b**) the relationship between key length and risk level

Given the above results, we can see that a more stable system will choose a more secure LMAC, since a more stable system is less sensitive to the delay.

Before studying the cyber strategy μ, we need to analyze the impact of detectors, defined by (11.8), on the POMDP problem. Since the True Negative (TN) rate of the detector only depends on $\bar{z}$, we first plot the relationship between TN rate and $\bar{z}$, illustrated in Fig. 11.10a. In Fig. 11.10a, we can see that the TN rate is increasing in $\bar{z}$. If the requirement of the TN rate is 0.9, then we need to choose $\bar{z} = 1.65$ according to Fig. 11.10a, i.e., $\phi_w(z_q) = 0.90$ and $\phi_w(z_u) = 0.10$.

The Missed Detection (MD) rate depends on the distance between fake reference $\bar{r}$ and real one r. To draw the ROC curve, we consider following four fake references: $\bar{r}_1 = [1.2, 0.1]^T$, $\bar{r}_2 = [1.3, 0.1]^T$, $\bar{r}_3 = [1.4, 0.2]^T$, $\bar{r}_4 = [1.5, 0.2]^T$. By moving $\bar{z}$ from 0 to 5, we plot the ROC curves based on the above fake references in Fig. 11.10b. According to Fig. 11.10b, we note that the distance between r and $\bar{r}$ is greater, the better the ROC curve will be, i.e., the better performance the detector will have. The reason is that a significant value of $\|r - \bar{r}\|_2$ will facilitate the detector to discover the abnormal situation. Given $\bar{r}_1$ and $\bar{r}_2$, we can see that the ROC curves

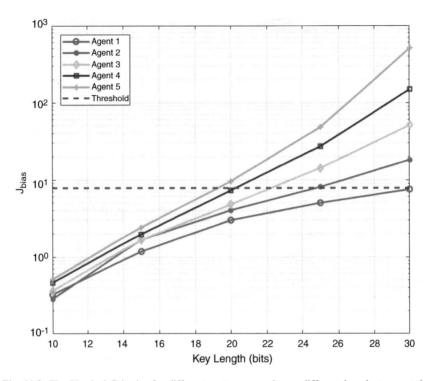

Fig. 11.9 The Physical Criteria: for different systems, we choose different lengths to meet the criteria

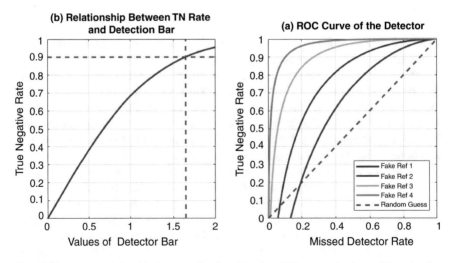

Fig. 11.10 (**a**) The relationship between the True Negative (TN) rate and values of detection bar $\bar{z}$; (**b**) ROC curve of the detector under different attack reference

are even below the curve of random guess when the MD rate is low; i.e., the detector cannot work in these regions.

In our experiment, we consider fake reference $\bar{r}_4$. Since we have $\phi_w(z_q) = 0.9$, we get $\phi_f(z_q) = 0.26$ based on the results in Fig. 11.10b. Therefore, the detection matrices $D(z)$, for $z \in \mathcal{Z}$, are given by

$$\Phi(z_q) = \begin{bmatrix} 0.90 & 0.00 \\ 0.00 & 0.26 \end{bmatrix}, \quad \Phi(z_u) = \begin{bmatrix} 0.10 & 0.00 \\ 0.00 & 0.74 \end{bmatrix}.$$

In the next part of the experiment, we will compute the cyber policy based on the above setups. Using the algorithm presented in Section V-D, we can compute the threshold ξ_i of cyber policy μ for each Agent i. Since we aim to see the interdependencies between physical parameters and cyber policy, we assume that all the ROS agents use the same key length $\ell_{key} = 18$ bits. The results of ξ_i are given by $\xi_1 = 0.7, \xi_2 = 0.76, \xi_3 = 0.82, \xi_4 = 0.88, \xi_5 = 0.94$. We can see that ξ_i is increasing in Agents' indexes. A significant value of ξ_i leads to a conservative strategy since ξ_i is the threshold, below which the agent will take the reboot action. According to the above results, we note that a more unstable agent will use a more conservative strategy. The reason is that a more unstable system will suffer a higher physical cost if the attacker compromises the lightweight MAC and launch a successful attack.

Figure 11.11 presents the Q functions, defined by (11.15), of different ROS agents. In Fig. 11.11, we can see that all the Q functions are non-increasing, and these results coincide with Theorem 11.3. To be more specific, the values of $Q(\cdot)$ remain the same with θ from 0 to ξ_i and start to decrease with θ from ξ_i to 1. The reason is that Agent i will take the reboot action when its belief is lower than threshold ξ_i. Secondly, we can see that for each $\theta \in \Theta$, the value of $Q(\cdot)$ is increasing in Agents' indices, since a more unstable system will lead to a higher physical cost.

The final experiment is to test the cyber performance of Agent 1 under the proposed resilient mechanism. A desirable cyber performance is a case that the detector will launch an alarm, and the agent can reboot immediately when the attacker compromises cyber defense. A poor cyber performance is a case that the agent fails to discover abnormal situations, or the detector launches too many false alarms. We have three case studies: the first one is a normal case with a desirable strategy ($\xi = 0.74, \bar{z} = 1.65$); the second one is a conservative case with strategy ($\xi = 0.92, \bar{z} = 1.40$); the third one is a radical case with strategy ($\xi = 0.65, \bar{z} = 2.80$). We aim to see the differences among these case studies.

Figure 11.12 illustrates the results of Case Study 1. In this case, we can see that detector succeeds in detecting a cyber attack at time $k = 6$, and the agent reboots itself to update its security mechanism. Even though the detector has a false alarm at time $k = 11$, the false alarm rate is negligible, and the performance is still satisfactory. Figure 11.13 presents the results of Case Study 2. In the conservative case, although the detector provides correct observations at time $k = 5$ and $k = 7$, it launches too many false alarms at time $k = 3, 5, 12, 14$. The conservative strategy

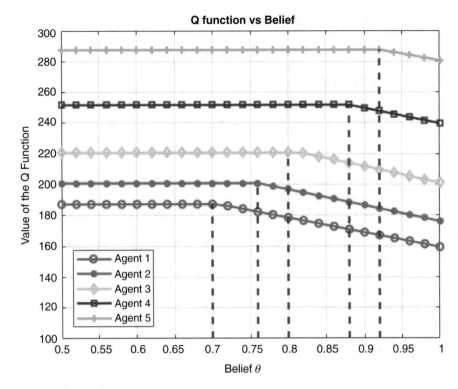

Fig. 11.11 The Q-functions defined by (11.15) of each agent: different agent will have different Q-functions and policies

keeps rebooting the system with high frequency, significantly lowing down the cyber performance. Figure 11.14 shows the results of Case Study 3. In the radical case, we can see that detector fails to present the alarm at $k = 6$, while the attacker compromises the security mechanism. The detector cannot provide correct information until $k = 9$. The radical strategy is dangerous since the attacker can inflict severe damage on the system from $k = 6$ to $k = 9$.

According to the above cyber experiments, we can see that the proposed resilient mechanism can protect the system from further damage after a successful attack. The desirable resilient strategy has high accuracy in detecting abnormal events.

11.5   Conclusions and Notes

In this chapter, we have designed a lightweight MAC (LMAC) for delay-sensitive ROS agents. Due to the risk of the LMAC and the incomplete information of the cyber events, we have formulated a cyber POMDP based on a Chi-square detector. We have developed a threshold policy to solve cyber POMDP. We have analyzed

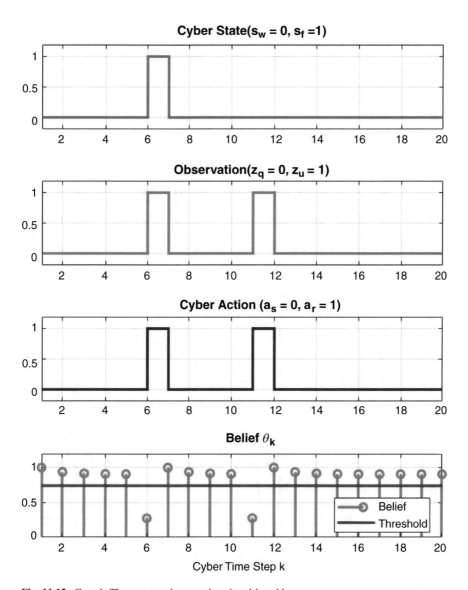

Fig. 11.12 Case 1: The system reboots at $k = 6$ and $k = 11$

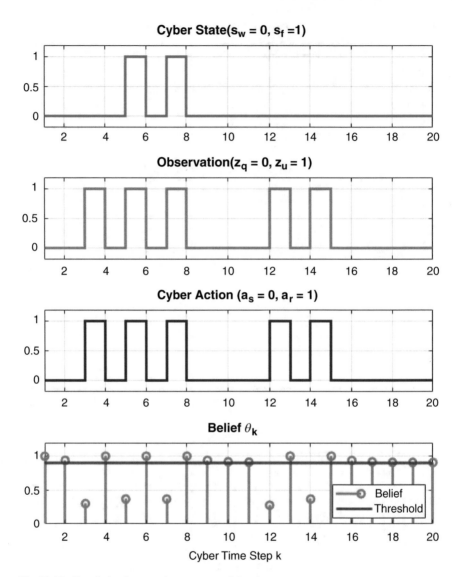

Fig. 11.13 Case 2: the detector gives too many false alarms

the relationship between the detection performance and the structural results of
the POMDP. Our numerical experiments have shown that a more unstable ROS
agent chooses a simpler LMAC and a more conservative resilient strategy since
it is more sensitive to the delay. These numerical results have indicated a strong
interdependency between the physical parameters and cyber policy.

This chapter has developed a co-design framework that integrates POMDPs as
a cyber layer model and a linear-quadratic control problem as a physical layer

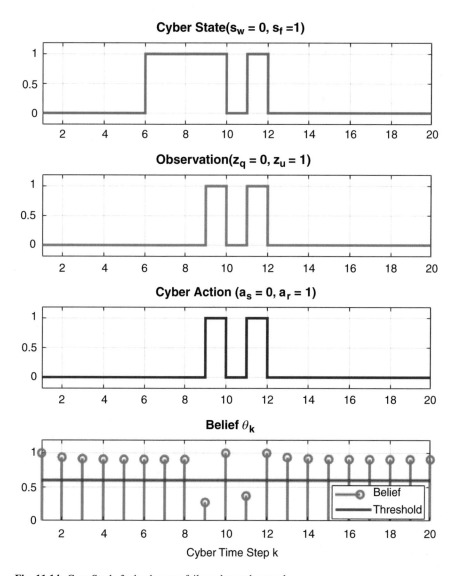

Fig. 11.14 Case Study 3: the detector fails to detect the attack

model. The design philosophy is similar to the ones that we have seen in Chaps. 7 and 8. The security of ROS is an emerging research topic. Recently, we have witnessed advances in securing ROS by securing the network [54, 79], deploying cryptographic solutions [41], and defending against deceptive behaviors [66]. It will be useful to incorporate adversarial models into the proposed framework in this chapter. This extension will naturally lead to game models, which provide essential theoretical underpinnings for designing defense strategies under adversarial environments.

Part IV
Discussion of the Future Work

Chapter 12
Future Work in Security Design of CPSs

12.1 Research Directions: Advanced Attack Models

This book has presented a cross-layer framework to design secure and resilient cyber-physical control systems. As we have seen in the previous chapters, the model that we choose for the cyber or the physical layer relies on the attack model and the control objectives of the CPSs. In particular, we have discussed in Part II cryptographic solutions to provide data confidentiality and integrity, and in Part III game-theoretic methods to enrich the attack models with the dimensions of incentives, resources, and information structures. In Chap. 7, we have used FlipIt game to capture the interactions between an attacker and a defender in an APT attack. In Chap. 8, we have used a zero-sum stochastic game model for the jamming attacks on CBTC systems, and in Chap. 9, a signaling game model for data integrity attacks in CPSs. In this section, we discuss two attack models that can be incorporated into the cross-layer framework as a future research direction.

12.1.1 Man-in-the-Middle Attack

In a Man-in-the-Middle (MITM) attack [145], the attackers send fake or false messages to the operators or agents. Figure 12.1 illustrates an example of the Man-in-the-Middle Attack in computer networks. The attacker first cuts off the original connection between the user and the web applications; then, it creates two new links to the user and the web application, respectively. During the attack, the attacker can stay stealthy and send fake data to both the users and the web application, misleading them to choose the wrong actions, e.g., leaking sensitive information.

© The Editor(s) (if applicable) and The Author(s), under exclusive license to
Springer Nature Switzerland AG 2020
Q. Zhu, Z. Xu, *Cross-Layer Design for Secure and Resilient Cyber-Physical Systems*,
Advances in Information Security 81, https://doi.org/10.1007/978-3-030-60251-2_12

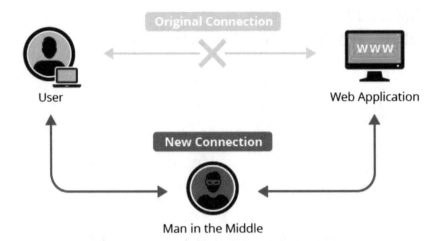

Fig. 12.1 An example of the Man-in-the-Middle Attack in computer applications: the attacker first cuts off the original connection between the user and web applications; then, it creates two new links to the user and the web application, respectively

The MITM attacks can create fatal losses to CPSs. For example, we consider a multi-agent Unmanned Aerial Vehicle (UAV) system with a control center that sends references to all the UAVs. If the attacker successfully launches a MITM attack, creating links with a certain number of UAVs. Then, the attacker can deviate the UAVs from their original trajectories by sending fake references. Even worse, the attacker can mislead the compromised UAVs to collide with other normal UAVs. By compromising a partial number of UAVs in the system, the attacker can damage the entire system based on a successful MITM attack.

12.1.2   Compromised-Key Attack

A key is a secret code that is necessary to interpret the secure information. Once an attacker obtains the secret key of the secure mechanism, we consider this attacks a Compromised-Key (C-Key) Attack [26]. Figure 12.2 illustrates the concept of the C-Key attack. If the attacker succeeds in a C-Key attack, then it can gain access to a secured communication without the perception of the sender or receiver by using the compromised key. The attacker can decrypt, modify the transmitting data with the keys. Besides, the attacker might use the compromised key to figure out the additional keys, allowing the attacker to access other communications channels. The C-Key attack can also have a significant on CPSs. For example, given the compromised key, the attacker can study the sensitive parameters of the by decrypting the transmitting data. Besides privacy attacks, the attacker can also launch a data-integrity attack by modifying the transmitting data. These two attacks will incur severe losses for CPSs.

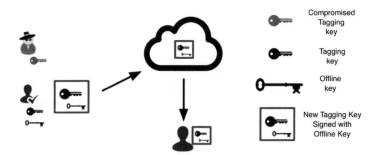

Fig. 12.2 An Compromised-Key Attack: the attacker first steal the secret key of the security mechanism; then it can use the compromised key to decrypt or modify the transmitting data

12.2 Research Directions: Data-Availability Issues in CPSs

In this book, we have focused on data confidentiality and integrity. In Chaps. 4 and 5, we have used cryptographic solutions to provide confidentiality and integrity to sensor and actuator signals. In Chaps. 9 and 11, we have used game-theoretic and POMDP approaches to address data integrity attacks on CPSs. Data availability plays an equally significant role in security issues of CPSs. A CPS has physical components and devices, which require real-time data coming from the cyber world. Attackers can disable material data by blocking cyber communications. If the physical system is a typical control system, then a data-availability attack can incur instability of the physical system since it requires feedback control inputs to guarantee stability at each sampling time.

Availability attacks are often easier to launch. For example, in a cloud-enabled CPS, the attacker can impede the communications between the CPS and cloud by sending false messages to the CPS [177]. Even though the CPS can verify the data, the massive false data will occupy all the buffer memory and make the data unavailable for the CPS. Hence one important research direction is to design mechanisms that can guarantee resiliency at the physical layer while assuring data availability at the cyber layer. In the following subsections, we discuss two potential solutions from both cyber and physical ends to mitigate the data-availability attack.

12.2.1 Safe-Mode Mechanism

One potential solution against data-availability attacks is to develop a safe-mode mechanism. Figure 12.3 illustrates the of the safe-mode tool based on an example of an automatic assembling line. In the safe-mode design, we develop a safe-mode controller that only require local data to realize essential functions and requirements (e.g., guaranteeing the stability) for the physical system.

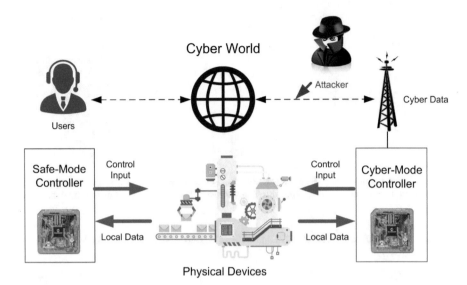

Fig. 12.3 The Safe-Mode Protection: We design a safe-mode controller for the physical system. When the attacker disables the cyber-mode controller by blocking the cyber communications, the physical system will activate the safe-mode controller, which only requires local data, to meet the requirements, e.g., stabilizing the physical system

Besides a safe-mode controller, the assembling line also has a cyber-mode controller, which can receive cyber data from remote users. In the proposed mechanism, if the attacker disables the cyber-mode controller by blocking the cyber communications, then the physical system will activate its safe-mode controller to guarantee the essential requirements, protecting the system.

12.2.2 Availability of a Partially Compromised System

Another potential solution against the data-availability attack is to ensure availability when the attacker only compromises some non-critical parts of the CPSs. In such situations, malicious attacks may not gain direct access to the critical components of CPSs. When the detector detects the compromised parts, the mechanism should be able to isolate the compromised parts while maintaining the availability of the CPSs. After separating from the compromised elements, the core parts of the CPSs can still realize essential functions.

12.3 Conclusions

In this final chapter, we have discussed several research directions that are motivated by advanced attack models. We could use our cross-layer design framework to couple the cyber solutions that address these attacks with the physical layer control system design. The overarching goal of this book is to introduce cross-layer designs as a design methodology for secure and resilient CPSs. One key challenge to attaining this objective is the need to bridge the gap between multiple disciplines. In this book, we have connected cryptography with control theory and bridged game theory with control theory to provide design methodologies for systems under different attack models. Another key challenge is the domain-specific knowledge to guide the CPS designs. We have used multiple case studies from UAVs, CBTC systems, cyber manufacturing, and robotic systems to illustrate the proposed methodologies. It is clear that each system has its own distinct features in terms of system constraints, specifications, and performance requirements. Hence domain knowledge plays an important role in selecting the right model for cyber defense and control design. Cross-layer design is not limited to the applications discussed in this book. On the contrary, it has broad applications in emerging areas such as autonomous driving systems, smart and connected health, sustainable energy and water systems. Security and resiliency will be a prominent concern in these systems. The cross-layer design paradigm would be a promising approach to address this concern.

Appendix A
Basics of Optimization

In this appendix, we review basic concepts from optimization theory. Consider a general optimization problem of the following form:

$$(\text{OP}) \quad \min_{x \in \mathcal{F}} f(x)$$

Here, one minimizes a function f over a set $\mathcal{F}$. Let $f : \mathbb{R}^n \to \mathbb{R}$ be a continuous real function of n variables defined on the entire space of $\mathbb{R}^n$. This function is called an objective function. Let $x \in \mathbb{R}^n$ be a point or a vector. Set $\mathcal{F} \subseteq \mathbb{R}^n$ is called the feasible set. The elements of $\mathcal{F}$ are feasible points. To compare distances between points, and talk about neighborhoods and local optima, we need the notion of a norm. We will work with the Euclidean norm, which is a function $\| \cdot \| : \mathbb{R}^n \to \mathbb{R}$ defined by $\|x\| = (x^T x)^{\frac{1}{2}}$. Here, x^T denotes the transpose of x.

Given two points $x, y \in \mathbb{R}^n$, the distance between x and y is $\|x - y\|$. A neighborhood of x^* is $\mathcal{N}(x^*) = \{x \in \mathbb{R}^n : \|x - x^*\| < \epsilon\}$ for some $\epsilon > 0$. In the the problem (OP), we look for a feasible point x that minimizes f over the set $\mathcal{F}$. We assume that $f(x)$ is finite at every given $x \in \mathcal{F}$. The optimization problems of the form (OP) are referred to as mathematical programs. A point $x^* \in \mathcal{F}$ is an optimal solution (or a global optimal solution or a global minimum) if $f(x^*) \leq f(x)$ for every $x \in \mathcal{F}$. We use the notation $x^* = \arg\min_{x \in \mathcal{F}} f(x)$. If $f(x^*) \leq f(x)$ for every $x \in \mathcal{F} \cap \mathcal{N}(x^*)$, where $\mathcal{N}(x^*)$ is some neighborhood of x^*, then x^* is a local optimal solution (or a local minimum).

For a given optimization problem, it is not necessary that an optimal solution exist. Suppose that there exists an $x^* \in \mathcal{F}$ such that $\inf\{f(x) : x \in \mathcal{F}\} = f(x^*)$. Then, we can write $f(x^*) = \inf_{x \in \mathcal{F}} f(x) = \min_{x \in \mathcal{F}} f(x)$. If such an x^* cannot be found, even though $\inf\{f(x) : x \in \mathcal{F}\}$ is finite, we say that an optimal solution does not exist. We call the quantity $\inf_{x \in \mathcal{F}} f(x)$ as the optimal value of the optimization

Q. Zhu, Z. Xu, *Cross-Layer Design for Secure and Resilient Cyber-Physical Systems*,
Advances in Information Security 81, https://doi.org/10.1007/978-3-030-60251-2

problem. If $\inf_{x \in \mathcal{F}} f(x)$ is not bounded from below, i.e., $\inf_{x \in \mathcal{F}} f(x) = -\infty$, then neither an optimal solution nor an optimal value exists.

In the problem (OP), an optimal solution exists if $\mathcal{F}$ is a finite set since there is only a finite number of comparisons to make. If $\mathcal{F}$ is not finite, the existence of an optimal solution is not always guaranteed. It is guaranteed if f is continuous and $\mathcal{F}$ is compact, a result known as the Weierstrass theorem. For the case where $\mathcal{F}$ is finite-dimensional, we should recall that compactness is equivalent to being closed and bounded.

A.1   Optimality Conditions for Unconstrained Problems

We say that f is differentiable at $x^* \in \mathbb{R}^n$ if there exists a row vector $r : \mathbb{R}^n \to \mathbb{R}$ such that

$$\lim_{h \to 0} \frac{1}{\|h\|} |f(x^* + h) - f(x^*) - r \cdot h| = 0.$$

The n-tuple r is called the derivative of f at x^*. It is denoted by $r = \nabla f(x^*)$. The function f is differentiable (or globally differentiable) if it is differentiable at every $x^* \in \mathbb{R}^n$, A differentiable function $f : \mathbb{R}^n \to \mathbb{R}$ is said to be continuously differentiable at x^* if a sequence $x^k \to x^*$ implies $\nabla f(x^k) \to \nabla f(x^*)$.

Consider an unconstrained problem of (OP), where $\mathcal{F} = \mathbb{R}^n$. Let f be a differential function. A first-order necessary condition for an optimal solution is to satisfy

$$\nabla f(x^*) = 0. \tag{A.1}$$

If, in addition, f is twice continuously differentiable on $\mathbb{R}^n$ on $\mathbb{R}^n$, a second-order necessary condition is

$$\nabla^2 f(x^*) \geq 0. \tag{A.2}$$

The pair of conditions (A1) and (A2) is sufficient for $x^* \in \mathcal{F}$ to be a locally minimizing solution. These conditions are also sufficient for global optimality if, in addition, $\mathcal{F}$ is a convex set and f is a convex function over $\mathcal{F}$.

A.2   Optimality Conditions for Constrained Problems

The main difficulty in the study of optimization problems (OP) is that the feasible set $\mathcal{F}$ is not given explicitly. In one type of formulations of optimization problems, this set is given in terms of equality constraints.

$$\text{(PE)} \qquad \min f(x)$$

$$h^i(x) = 0, \ \ i \in \mathcal{P}$$

Here, $h^i : \mathbb{R}^n \to \mathbb{R}, i \in \mathcal{P}$, are continuous functions, and $\mathcal{P} = \{1, 2, \cdots, m\}$ is a finite index set. The optimization problems (PE) have been studied by Lagrange and, in abstract spaces, by Euler. They have formulated a condition that identifies a subset of the feasible set $\mathcal{F} = \{x : h^i(x) = 0, i \in \mathcal{P}\}$ that contains all locally optimal solutions.

Theorem A.1 (Method of Lagrange) *Consider the program (PE), where the objective function f is assumed to be differentiable and the constraints $h^i, i \in \mathcal{P}$, are assumed to be continuously differentiable at a feasible point x^*. The necessary condition for optimality is that if x^* is a local optimum, then the gradients*

$$\nabla f(x^*), \ \nabla h^1(x^*), \cdots, \nabla h^m(x^*)$$

are linearly dependent. It is known as the Euler-Lagrange condition for optimality.

If the gradients of the m constraints are linearly independent, then the gradient of the objective function can be expressed as a linear combination of these gradients, i.e., the system

$$\nabla f(x^*) + \sum_{i \in \mathcal{P}} \lambda_i \nabla h^i(x^*) = 0$$

is consistent (for some $\lambda_i \in \mathbb{R}, i \in \mathcal{P}$). This observation leads to the following well-known method of Lagrange.

Theorem A.2 *Consider the program (PE). Assuming that the gradients of the constraints are linearly independent at an optimal solution, the method of Lagrange consists of solving the system of $m + n$ equations:*

$$\nabla f(x) + \sum_{i \in \mathcal{P}} \lambda_i \nabla h^i(x) = 0 \tag{A.3}$$

$$h^i(x) = 0, i \in \mathcal{P} \tag{A.4}$$

in $m + n$ unknowns $x \in \mathbb{R}^n$ and $x = (\lambda_i) \in \mathbb{R}^m$. If x^* and λ^* solve this system, then we say that x^* is a candidate for local optimality.

The m scalars $\lambda_i, i \in \mathcal{P}$, are called Lagrange multipliers. They arise in the dual problems, and they are important for sensitivity analyses and economic interpretations. The method of Lagrange gives necessary conditions for optimality. To establish whether a point x^* obtained by the method is indeed optimal, one has to apply a sufficient condition for optimality. These conditions typically require the second derivative, in which case they are referred to as "second-order optimality condition." The second derivation can also be used to strengthen a first-order necessary condition for optimality by identifying feasible points as non-optimal.

Consider the optimization problem (OP) where the feasible set is determined by both equality and inequality constraints:

$$(NP) \qquad \min f(x)$$

$$h^i(x) = 0, \ i \in \mathcal{P}$$

$$g^j(x) \le 0, \ j \in \mathcal{Q}.$$

Here, $f, h^i, g^j : \mathbb{R}^n \to \mathbb{R}$ are continuous functions, $i \in \mathcal{P}, j \in \mathcal{Q}$, and $\mathcal{P} = \{1, 2, \cdots, p\}$ and $\mathcal{Q} = \{1, 2, \cdots, q\}$ are finite index sets. The feasible set of (NP) is $\mathcal{F} := \{x \in \mathbb{R}^n : h^i(x) = 0, g^j(x) \le 0, \ i \in \mathcal{P}, \ j \in \mathcal{Q}\}$. Given a feasible point $x^* \in \mathcal{F}$, we want to check whether the point is locally optimal. First we determine the index set of active inequality constraints at x^* defined as $\mathcal{Q}(x^*) = \{j \in \mathcal{Q} : g^j(x^*) = 0\}$. The idea of introducing $\mathcal{Q}(x^*)$ is to identify the constraints that matter in the description of local optimality of the point x^*. Note that every index in $\mathcal{P}$, the set of equality constraints, is active at every feasible point. Introduce the set

$$D(x^*) :=$$

$$\{d \in \mathbb{R}^n : \nabla f(x^*)d \le 0, \ \nabla h^i(x^*)d = 0, \ i \in \mathcal{P}, \ \nabla g^j(x^*)d \le 0, \ j \in \mathcal{Q}(x^*)\}$$

and the Lagrangian function

$$L(x, \mu, \lambda) = \mu_0 f(x) + \sum_{i \in \mathcal{P}} \lambda_i h^i(x) + \sum_{j \in \mathcal{Q}} \mu_j g^j(x).$$

Note that $D(x^*)$ is a nonempty set. The leading coefficient μ_0 is associated with the objective function. This coefficient may be zero in some special situations. The first and the second derivatives of L, relative to x, are denoted by $\nabla_x L$ and $\nabla_x^2 L$, respectively. Given any $x^* \in \mathcal{F}$, one can use the following second-order necessary condition for optimality to determine whether x^* could be a local optimum.

Theorem A.3 (Second-Order Necessary Condition for Local Minimum) *Consider the program (NP) where all functions are assumed to be twice continuously differentiable at a feasible point x^*. If x^* is a locally optimal solution, then, for every $d \in D(x^*)$, there exist multipliers $\mu_0 \ge 0, \mu_j \ge 0, j \in \mathcal{Q}$, and $\lambda_i \in \mathbb{R}$, $i \in \mathcal{P}$, not all zero, such that*

$$\nabla_x L(x^*, \mu, \lambda) = 0 \qquad (A.5)$$

$$d^T \nabla_x^2 L(x^*, \mu, \lambda)d \ge 0 \qquad (A.6)$$

$$\mu_j g^j(x^*) = 0, j \in \mathcal{Q} \qquad (A.7)$$

$$\mu_0 \nabla f(x^*)d = 0 \qquad (A.8)$$

$$\mu_j \nabla g^j(x^*)d = 0, \quad j \in Q \tag{A.9}$$

Note that the multipliers in the theorem depend on the choice of d. The q equations in (A.7) are called the "complementary conditions." If a feasible x^* satisfies the above necessary conditions, then x^* still may not be optimal. Its optimality can be checked by a sufficient condition for optimality. The condition given below checks whether x^* is an isolated local minima. These are local minima $x^* \in \mathcal{F}$ with the additional property that $f(x^*) < f(x)$ for every $x \in \mathcal{F} \cap \mathcal{N}(x^*), x \neq x^*$, where $\mathcal{N}(x^*)$ is some neighborhood of x^*.

Theorem A.4 (Second-Order Sufficient Condition for Isolated Local Minimum) *Consider the program (NP) where all functions are assumed to be twice continuously differentiable. A feasible point x^* is an isolated local minimum if either $D(x^*) = \{0\}$ or if, for every $d \in D(x^*), d \neq 0$, there exist multipliers $\mu_0 \geq 0$, $\mu_j \geq 0, j \in Q$, and $\lambda_i \in \mathbb{R}, i \in \mathcal{P}$, not all zero, such that*

$$\nabla_x L(x^*, \mu, \lambda) = 0 \tag{A.10}$$

$$d^T \nabla_x^2 L(x^*, \mu, \lambda)d > 0 \tag{A.11}$$

$$\mu_j g^j(x^*) = 0, j \in Q \tag{A.12}$$

$$\mu_0 \nabla f(x^*)d = 0 \tag{A.13}$$

$$\mu_j \nabla g^j(x^*)d = 0, \quad j \in Q \tag{A.14}$$

The main difference between the second-order necessary and the second-order sufficient condition for optimality. The inequality (A.6) in Theorem A.3 is required to hold strictly in Theorem A.4. The conditions in Theorems A.3 and A.4 are often difficult to verify. Readers can refer to [17] for algorithmic solutions to general optimization problems. There are many extensions and results of optimality conditions for optimization problems for special classes of optimization problems. See [17, 98, 213] for further results.

Appendix B
Basics of Linear-Quadratic Optimal Control

In this appendix, we give an overview of the linear-quadratic optimal control theory. The term "linear-quadratic" refers to the linear system dynamics and the quadratic cost function. Linear-quadratic regulator (LQR) is a particular case of optimal control, specialized for linear system dynamics and quadratic cost-function. The optimal control problem in general has two different formulations: finite time and infinite horizon.

B.1 Finite-Time Optimal Control Problem Formulation

For an LTI system

$$\dot{x}(t) = Ax(t) + Bu(t), \quad x(t_0) = x_0, \tag{B.1}$$

where $x(t), x_0 \in \mathbb{R}^n$, $A \in \mathbb{R}^{n \times n}$, $B \in \mathbb{R}^{n \times m}$, $u(t) \in \mathbb{R}^m$, find a control law $u(\cdot)$ that minimizes the quadratic cost function

$$
\begin{aligned}
& J\left(x(t_0), u(\cdot), t_0, T\right) \\
& = \underbrace{x^\top(T)Mx(T)}_{m(x(T))} + \int_{t_0}^{T} \underbrace{\left(x^\top(t)Qx(t) + u^\top(t)Ru(t)\right)}_{f(x,u,t)} dt \to \min
\end{aligned}
\tag{B.2}
$$

where $M \in \mathbb{R}^{n \times n}$ and $Q \in \mathbb{R}^{n \times n}$ are positive semidefinite matrices, $R \in \mathbb{R}^{m \times m}$ is a positive definite matrix, T is fixed, while $x(T) \in \mathbb{R}^n$ is free.

The finite-time optimal control problem formulation can otherwise be stated as follows: find a control law that takes the system from the point x_0 to "another

Q. Zhu, Z. Xu, *Cross-Layer Design for Secure and Resilient Cyber-Physical Systems*,
Advances in Information Security 81, https://doi.org/10.1007/978-3-030-60251-2

point $x(T)$" in the given finite time T, while minimizing the given cost-function (which can be energy or fuel consumption, for example). Notice that once the minimizing control law u is found for the given time interval $[t_0, T]$, $x(T)$ will be determined from the unique solution of the dynamical system $\dot{x}(t) = Ax(t) + Bu(t)$. Therefore it was important to mention that $x(T)$ is free. The optimal control problem formulation can also be given for the finite time T being free, including $T \to \infty$. Similarly, constraints can be imposed on initial or boundary conditions. Within the limits of this course, we will consider the finite-time optimal problem formulation above and its infinite-horizon counterpart.

We see that the cost function has a boundary condition and also two terms in the integrand. It is important to understand the physical meaning of these two terms. The first one $x^\top(t)Qx(t)$ penalizes the tracking error, while the second one $u^\top(t)Ru(t)$ penalizes the control effort. Since both terms are nonnegative definite, then obviously minimization of the cost function in the presence of large Q will lead to smaller tracking errors, and in the presence of large R will lead to smaller control efforts. Since "smaller control efforts" may fail to achieve the control objective, then it becomes obvious that the selection of Q and R should be traded-off cleverly. The tuning of Q and R is the main challenge in the design of the optimal controller. The solution to this problem is given via $u(t) = -K(t)x(t)$, where the computation of $K(t)$ involves a *differential Riccati equation*, dependent upon Q and R, in addition to the system matrices.

B.2   Infinite Horizon Optimal Control Problem Formulation

For a stabilizable LTI system

$$\dot{x}(t) = Ax(t) + Bu(t), \quad x(t_0) = x_0, \tag{B.3}$$

find a control law $u(\cdot)$ that stabilizes the origin (i.e., *regulates* x to zero), while minimizing the quadratic cost function

$$J(x(t_0), u(\cdot), t_0) = \int_{t_0}^{\infty} \left(x^\top(t)Qx(t) + u^\top(t)Ru(t) \right) dt \tag{B.4}$$

where $Q = Q^\top \geq 0$ and $R = R^\top > 0$.

The solution to this problem is given via $u(t) = -Kx(t)$, where the computation of K involves an *algebraic Riccati equation*. Notice that the infinite time problem formulation aims at regulation of the system state to the origin ($x|_{T \to \infty} = 0$), because of which the cost function does not have the additive term dependent upon the final boundary condition.

B.3   Principle of Optimality

Before giving the derivations for the Riccati equations, we need to state the *principle of optimality*: If $u^*(\tau)$ is optimal over the interval $[t, T]$, then $u^*(\tau)$ is necessarily optimal over the subinterval $[t + \Delta t, T]$ for any Δt such that $T - t \geq \Delta t > 0$.

Hamilton-Jacobi-Bellman-Isaacs equation is the fundamental equation in optimal control, which defines the optimal control structure via the solution of a partial differential equation. We will derive it now for the finite-time problem formulation, and then consider the limiting case as $T \to \infty$.

From Eq. (B.2), we see that the performance index J depends on the initial state $x(t_0)$ and time t_0, the control $u(\cdot)$ and the final time T. However, rather than just searching for the control $u(\cdot)$ minimizing (B.2) for various $x(t_0)$, next we shall study the determination of the optimal control for all initial conditions t and $x(t)$, that is the control law $u(\cdot)$ which minimizes:

$$J\left(x(t), u(\cdot), t, T\right) = \underbrace{x^\top(T)Mx(T)}_{m(x(T))} + \int_t^T \underbrace{\left(x^\top(t)Qx(t) + u^\top(t)Ru(t)\right)}_{f(x,u,t)} dt \,.$$

(B.5)

Then, if the system starts in state $x(t)$ at time t, the minimal value of the cost-function with the minimization being done over $u(\cdot)$ defined on $[t, T]$ is given by:

$$J^*(x(t), t, T) = \min_{u(\cdot)} \left(\int_t^T f(x, u, \tau)d\tau + m(x(T)) \right) \,.$$

Notice that $J^*(x(t), t, T)$ is independent of $u(\cdot)$, precisely because knowledge of the initial state and time abstractly determines the particular control, by the requirement that the control minimizes $J\left(x(t), u(\cdot), t, T\right)$.

From the additive properties of the integral and the optimality principle we have that:

$$J^*(x(t), t, T) = \min_{u(\cdot)} \left(\int_t^{t+\Delta t} f(x, u, \tau)d\tau + J^*\left(x(t + \Delta t), t + \Delta t, T\right) \right) \,,$$

where we have used the fact that the minimal value of J starting in state $x(t + \Delta t)$ at time $t + \Delta t$ is given by $J^*\left(x(t + \Delta t), t + \Delta t, T\right)$. Thus finding the minimum over $[t, T]$, due to the optimality principle, is reduced to finding the minimum over the reduced interval $[t, t + \Delta t]$. Taking arbitrarily small Δt, we can approximate the integral $\int_t^{t+\Delta t} f(x, u, \tau)d\tau$ by $f(x, u, t)\Delta t$, and also expand $J^*\left(x(t + \Delta t), t + \Delta t, T\right)$ into Taylor series around $(x(t), t)$ to obtain:

$$J^*(x, t, T)$$

$$= \min_u \left(f(x, u, t)\Delta t + J^*(x, t, T) + \frac{\partial J^*}{\partial t}\Delta t + \left(\frac{\partial J^*}{\partial x}\right)^\top \dot{x}(t)\Delta t + o(\Delta t) \right),$$

where $\frac{\partial J^*}{\partial x}$ is the gradient of J^* with respect to the vector x, while $o(\Delta t)$ denotes higher order terms. Canceling out $J^*(x, t, T)$ from left and right hand sides, dividing by Δt, and taking the limit as $\Delta t \to 0$, we obtain

$$-\frac{\partial J^*}{\partial t} = \min_u \mathcal{H}\left(t, x, u, \frac{\partial J^*}{\partial x}\right), \tag{B.6}$$

where

$$\mathcal{H}\left(t, x, u, \frac{\partial J^*}{\partial x}\right) = f(x, u, t) + \left(\frac{\partial J^*}{\partial x}\right)^\top \dot{x}(t)$$

is called *Hamiltonian* of the system. Meantime

$$J^*(x(T)) = m(x(T)) \tag{B.7}$$

defines the boundary condition for the solution of the PDE. Thus, we obtained the partial differential equation (B.6) and the boundary condition (B.7) for the optimal value of $J^*(x, t, T)$. We need to find the *optimal control* $u^*(t)$ that achieves the minimum. The equation in (B.6) is known as Hamilton-Jacobi-Bellman-Isaacs (HJBI) equation, while $\mathcal{H}\left(t, x, u, \frac{\partial J^*}{\partial x}\right)$ is referred to as Hamiltonian. It is important to notice that the HJBI equation was derived, assuming that $J^*(x, t, T)$ is the optimal value of the cost function. Thus, the HJBI equation is a *necessary condition for optimality*. In fact, one can prove that under sufficient smoothness assumptions it is also a *sufficient condition*, i.e., if there exists a smooth cost-function $J'(x, t, T)$ that satisfies the HJBI equation, then it is also a minimum cost-function, $J'(x, t, T) = J^*(x, t, T)$.

B.4   Finite-Time Linear-Quadratic Optimal Control

We use the linear system dynamics and the quadratic cost-function to obtain the resulting control law explicitly. The PDE takes the form:

$$-\frac{\partial J^*}{\partial t} = \min_u \left(x^\top(t)Qx(t) + u^\top(t)Ru(t) + \left(\frac{\partial J^*}{\partial x}\right)^\top (Ax + Bu) \right).$$

To find the minimum, we need to take the derivative of the right-hand side with respect to $u(\cdot)$ and set it equal to zero to solve:

$$u^* = -\frac{1}{2}R^{-1}B^\top\frac{\partial J^*}{\partial x} \tag{B.8}$$

We need to substitute this back into the equation above to see what we get:

$$-\frac{\partial J^*}{\partial t} = x^\top(t)Qx(t) + \frac{1}{4}\left(\frac{\partial J^*}{\partial x}\right)^\top B(R^{-1})^\top B^\top\frac{\partial J^*}{\partial x}$$

$$+\left(\frac{\partial J^*}{\partial x}\right)^\top\left(Ax - \frac{1}{2}BR^{-1}B^\top\frac{\partial J^*}{\partial x}\right).$$

Assume that the optimal value of the cost-function has the following structure

$$J^*(x, t, T) = x^\top(t)P(t, T)x(t), \tag{B.9}$$

where $P(t, T)$ is symmetric. We notice that this assumed structure has T in the argument of $P(t, T)$. For the sake of simplicity, we will use the notation $P(t)$ instead of $P(t, T)$, but we have to keep in mind this dependence. Then direct substitution implies:

$$-\dot{x}^\top(t)P(t)x(t) - x^\top(t)\dot{P}(t)x(t) - x^\top(t)P(t)\dot{x}(t)$$

$$= x^\top(t)Qx(t) + \frac{1}{4}(2P(t)x(t))^\top B(R^{-1})^\top B^\top(2P(t)x(t))$$

$$+ (2P(t)x(t))^\top\left(Ax(t) - \frac{1}{2}BR^{-1}B^\top(2P(t)x(t))\right).$$

We rewrite

$$-(Ax(t) - BR^{-1}B^\top P(t)x(t))^\top P(t)x(t) - x^\top(t)\dot{P}(t)x(t)$$

$$-x^\top(t)P(t)(Ax(t) - BR^{-1}B^\top P(t)x(t))$$

$$= x^\top(t)Qx(t) + x^\top(t)P^\top(t)B(R^{-1})^\top B^\top P(t)x(t)$$

$$+2x^\top(t)P^\top(t)(Ax(t) - BR^{-1}B^\top P(t)x(t)).$$

Taking into consideration that $P(t)$ is symmetric, we have $2x^\top(t)P(t)Ax(t) = x^\top(t)(A^\top P(t) + P(t)A)x(t)$, which consequently reduces the above equation to the following structure:

$$-x^\top(t)\dot{P}(t)x(t) = x^\top(t)(A^\top P(t) + P(t)A + Q - P(t)BR^{-1}B^\top P(t))x(t),$$

which holds if and only if $P(t)$ satisfies the following *differential Riccati equation*:

$$\dot{P}(t) + A^\top P(t) + P(t)A + Q - P(t)BR^{-1}B^\top P(t) = 0 \tag{B.10}$$

along with the boundary condition:

$$J^*(x(t), t, T)\Big|_{t=T} = J^*(x(T), T, T)$$

$$= x^\top(T)P(t, T)\Big|_{t=T} x(T) = x^\top(T)Mx(T),$$

which immediately gives the boundary condition in the following form:

$$P(t, T)\Big|_{t=T} = P(T) = M. \tag{B.11}$$

Thus, if the differential Riccati equation (B.10)–(B.11) has a symmetric solution for $P(t)$, then the optimal control is given by

$$u^*(t) = -K(t)x(t), \tag{B.12}$$

where

$$K(t) = R^{-1}B^\top P(t). \tag{B.13}$$

Notice that the assumed structure for $J^*(x, t, T) = x^\top(t)P(t, T)x(t)$ along with the optimal control definition in (B.8) verifies the condition of the minimum:

$$x^\top(t)P(t, T)x(t)$$

$$= \int_t^T \left(x^\top(\tau)Qx(\tau) + x^\top(\tau)P^\top(\tau)B(R^{-1})^\top B^\top P(\tau, T)x(\tau) \right)d\tau$$

$$+ x^\top(T)Mx(T).$$

Indeed, taking the derivative of both sides and using the Riccati equation, one obtains trivial equality. Thus, we arrive at the following result.

Lemma B.1 *If $P(t)$ solves the Riccati equation in (B.10) along with the boundary condition in (B.11), then the controller in (B.12)–(B.13) achieves the objective of optimal control problem formulation for the system in (B.1) and cost function in (B.2) for the given finite time interval $[t_0, T]$.*

Note that we considered an LTI system, and the optimal control solution ended up having time-varying gain. This was due to the nature of the optimal control problem formulation, which was defined for *finite time interval*. On the offset, it may appear that this was due to the fact that we looked for the solution in a specific structure $J^*(x, t, T) = x^\top P(t, T)x$, and one might question the need for the time-dependence in P. If we had considered time-invariant P in this structure, then there would be no opportunity to verify the terminal condition, as $J^*(x, T, T) = x^\top Mx$ would immediately imply that $P = M$, which in its turn may not solve the resulting algebraic Riccati equation. As we show below, for the infinite-horizon case, when the cost-function does not have the terminal term $x^\top(T)Mx(T)$, we will obtain the algebraic Riccati equation, implying that P is a constant matrix, which in its turn will define an LTI *optimal* controller for an LTI system.

B.5   Infinite-Time Linear-Quadratic Optimal Control

As compared to the finite-time optimal control case, we need to establish that as $T \to \infty$, the Riccati equation has a well-defined solution, and the resulting controller stabilizes the system.

Theorem B.1 *Given the system in (B.3) and the cost-function in (B.4) with $R > 0$, and $Q = C^\top C$, where the pair (A, C) is detectable and the pair (A, B) is stabilizable, the optimal stabilizing controller is a linear constant gain feedback*

$$u^*(t) = -Kx(t),$$

where

$$K = R^{-1}B^\top\bar{P},$$

and $P = P^\top > 0$ is the unique symmetric and positive definite solution of the algebraic Riccati equation

$$A^\top\bar{P} + \bar{P}A + Q - \bar{P}BR^{-1}B^\top\bar{P} = 0. \tag{B.14}$$

Furthermore, $\bar{P}$ is the limit for $T \to \infty$ of the solution of the differential Riccati equation (B.10) with final condition $P(T, T) = 0$, that is

$$\bar{P} = \lim_{T\to\infty} P(t, T). \tag{B.15}$$

This appendix gives an overview of basic results of linear-quadratic optimal control. For more a more comprehensive introduction to control theory, see [9, 29, 183].

References

1. M.A. Abdelrazik, J.F. Hodapp, The E-SAT 300A: a multichannel satellite communication system for aircraft, in *GLOBECOM 1989* (IEEE, New York, 1989), pp. 1423–1427
2. F. Abdi, C.-Y. Chen, M. Hasan, S. Liu, S. Mohan, M. Caccamo, Guaranteed physical security with restart-based design for cyber-physical systems, in *Proceedings of the 9th ACM/IEEE International Conference on Cyber-Physical Systems* (IEEE Press, New York, 2018), pp. 10–21
3. H. Abou-Kandil, G. Freiling, G. Jank, On the solution of discrete-time markovian jump linear quadratic control problems. Automatica **31**(5), 765–768 (1995)
4. Additive vs Subtractive manufacturing. http://www.approto.com. Last accessed in 25 Aug 2015
5. E. Al-Shaer, Toward network configuration randomization for moving target defense, in *Moving Target Defense* (Springer, New York, 2011), pp. 153–159
6. E. Altman, K. Avratchenkov, N. Bonneau, M. Debbah, R. El-Azouzi, D. S. Menasché, Constrained stochastic games in wireless networks, in *IEEE GLOBECOM 2007-IEEE Global Telecommunications Conference* (IEEE, New York, 2007), pp. 315–320
7. M. Arnold, G. Andersson, Model predictive control of energy storage including uncertain forecasts, in *Power Systems Computation Conference (PSCC)*, Stockholm (2011)
8. R. Arumugam, V. Enti, L. Bingbing, W. Xiaojun, K. Baskaran, F. Kong, A. Kumar, K. Meng, G. Kit, DAvinCi: a cloud computing framework for service robots, in *IEEE International Conference on Robotics and Automation (ICRA)* (2010), pp. 3084–3089
9. M. Athans, P.L. Falb, *Optimal Control: An Introduction to the Theory and Its Applications* (Courier Corporation, Chelmsford, 2013)
10. S. Backhaus, R. Bent, J. Bono, R. Lee, B. Tracey, D. Wolpert, D. Xie, Y. Yildiz, Cyber-physical security: a game theory model of humans interacting over control systems. IEEE Trans. Smart Grid **4**(4), 2320–2327 (2013)
11. T. Başar, P. Bernhard, *H-infinity Optimal Control and Related Minimax Design Problems: A Dynamic Game Approach* (Springer Science & Business Media, Berlin/Heidelberg, 2008), pp. 3084–3089
12. T. Başar, G.J. Olsder, *Dynamic Noncooperative Game Theory*, vol. 200 (SIAM, Philadelphia, PA, 1995)
13. T. Basar, G.J. Olsder, *Dynamic Noncooperative Game Theory*, vol. 23 (SIAM, Philadelphia, PA, 1999)

Q. Zhu, Z. Xu, *Cross-Layer Design for Secure and Resilient Cyber-Physical Systems*,
Advances in Information Security 81, https://doi.org/10.1007/978-3-030-60251-2

14. M. Bellare, J. Kilian, P. Rogaway, The security of the cipher block chaining message authentication code. J. Comput. Syst. Sci. **61**(3), 362–399 (2000)

15. V. Bernardo, M. Curado, T. Staub, T. Braun, Towards energy consumption measurement in a cloud computing wireless testbed, in *2011 First International Symposium on Network Cloud Computing and Applications* (IEEE, New York, 2011), pp. 91–98

16. D.P. Bertsekas, *Dynamic Programming and Optimal Control*, vol. 1 (Athena Scientific, Belmont, MA, 1995)

17. D.P. Bertsekas, *Nonlinear Optimization* (Athena Scientific, Belmont, 1999)

18. B. Bielefeldt, J. Hochhalter, D. Hartl, Computationally efficient analysis of sma sensory particles embedded in complex aerostructures using a substructure approach, in *ASME 2015 Conference on Smart Materials, Adaptive Structures and Intelligent Systems*. American Society of Mechanical Engineers Digital Collection (2016)

19. S. Bitam, A. Mellouk, ITS-cloud: cloud computing for intelligent transportation system, in *2012 IEEE Global Communications Conference (GLOBECOM)* (IEEE, New York, 2012), pp. 2054–2059

20. M.Z. Bjelica, B. Mrazovac, V. Vojnovic, I. Papp, Gateway device for energy-saving cloud-enabled smart homes, in *2012 Proceedings of the 35th International Convention MIPRO* (IEEE, New York, 2012), pp. 865–868

21. S. Boyd, L. Vandenberghe, *Convex Optimization* (Cambridge University Press, Cambridge, 2009)

22. G. Cai, B.M. Chen, X. Dong, T.H. Lee, Design and implementation of a robust and nonlinear flight control system for an unmanned helicopter. Mechatronics **21**(5), 803–820 (2011)

23. E.F. Camacho, C.B. Alba, *Model Predictive Control* (Springer Science & Business Media, Berlin/Heidelberg, 2013)

24. W.A. Casey, Q. Zhu, J.A. Morales, B. Mishra, Compliance control: managed vulnerability surface in social-technological systems via signaling games, in *Proceedings of the 7th ACM CCS International Workshop on Managing Insider Security Threats* (2015), pp. 53–62

25. A. Cetinkaya, H. Ishii, T. Hayakawa, Event-triggered output feedback control resilient against jamming attacks and random packet losses. IFAC-PapersOnLine **48**(22), 270–275 (2015)

26. K. Chalkias, F. Baldimtsi, D. Hristu-Varsakelis, G. Stephanides, Two types of key-compromise impersonation attacks against one-pass key establishment protocols, in *International Conference on E-Business and Telecommunications* (Springer, New York, 2007), pp. 227–238

27. Y.H. Chang, Q. Hu, C.J. Tomlin, Secure estimation based kalman filter for cyber–physical systems against sensor attacks. Automatica **95**, 399–412 (2018)

28. M. Cheminod, L. Durante, A. Valenzano, Review of security issues in industrial networks. IEEE Trans. Ind. Inf. **9**(1), 277–293 (2013)

29. C.-T. Chen, *Introduction to Linear System Theory* (Holt, Rinehart and Winston, New York, 1970)

30. J. Chen, Q. Zhu, Interdependent network formation games with an application to critical infrastructures, in *2016 American Control Conference (ACC)*, pp. 2870–2875 (IEEE, New York, 2016)

31. J. Chen, Q. Zhu, Resilient and decentralized control of multi-level cooperative mobile networks to maintain connectivity under adversarial environment, in *2016 IEEE 55th Conference on Decision and Control (CDC)* (IEEE, New York, 2016), pp. 5183–5188

32. J. Chen, Q. Zhu, Control of multi-layer mobile autonomous systems in adversarial environments: a games-in-games approach. IEEE Trans. Control Netw. Syst. (2019). arXiv:1912.04082v

33. J. Chen, C. Touati, Q. Zhu, Heterogeneous multi-layer adversarial network design for the IoT-enabled infrastructures, in *GLOBECOM 2017-2017 IEEE Global Communications Conference* (IEEE, New York, 2017), pp. 1–6

34. Y. Chen, S. Kar, J.M. Moura, Optimal attack strategies subject to detection constraints against cyber-physical systems. IEEE Trans. Control Netw. Syst. **5**(3), 1157–1168 (2017)

35. L.G. Cima, M.J. Cima, Computer aided design; stereolithography, selective laser sintering, three-dimensional printing, Feb. 13 (1996). US Patent 5,490,962

36. G. Como, B. Bernhardsson, A. Rantzer, *Information and Control in Networks*, vol. 450 (Springer, New York, 2013)

37. O.L.V. Costa, M.D. Fragoso, R.P. Marques, *Discrete-Time Markov Jump Linear Systems* (Springer Science & Business Media, Berlin/Heidelberg, 2006)

38. P.V. Craven, A brief look at railroad communication vulnerabilities, in *The 7th International IEEE Conference on Intelligent Transportation Systems, 2004. Proceedings* (IEEE, New York, 2004), pp. 245–249

39. M.K. Daly, Advanced persistent threat. Usenix, Nov **4**(4), 2013–2016 (2009)

40. S.M. Dibaji, M. Pirani, D.B. Flamholz, A.M. Annaswamy, K.H. Johansson, A. Chakrabortty, A systems and control perspective of cps security. Annu. Rev. Control **47**, 394–411 (2019)

41. B. Dieber, S. Kacianka, S. Rass, P. Schartner, Application-level security for ROS-based applications, in *2016 IEEE/RSJ International Conference on Intelligent Robots and Systems (IROS)* (IEEE, New York, 2016), pp. 4477–4482

42. G. Dimitrakopoulos, P. Demestichas, Intelligent transportation systems. IEEE Veh. Technol. Mag. **5**(1), 77–84 (2010)

43. D.V. Dimarogonas, E. Frazzoli, K.H. Johansson, Distributed event-triggered control for multi-agent systems. IEEE Trans. Autom. Control **57**(5), 1291–1297 (2011)

44. D. Ding, Q.-L. Han, Y. Xiang, X. Ge, X.-M. Zhang, A survey on security control and attack detection for industrial cyber-physical systems. Neurocomputing **275**, 1674–1683 (2018)

45. A.W. Evans, Fatal train accidents on europe's railways: 1980–2009. Accid. Anal. Prev. **43**(1), 391–401 (2011)

46. M.J. Farooq, Q. Zhu, Modeling, analysis, and mitigation of dynamic botnet formation in wireless IoT networks. IEEE Trans. Inf. Foren. Secur. **14**(9), 2412–2426 (2019)

47. M.J. Farooq, H. ElSawy, Q. Zhu, M.-S. Alouini, Optimizing mission critical data dissemination in massive IoT networks, in *2017 15th International Symposium on Modeling and Optimization in Mobile, Ad Hoc, and Wireless Networks (WiOpt)* (IEEE, New York, 2017), pp. 1–6

48. H. Fawzi, P. Tabuada, S. Diggavi, Secure estimation and control for cyber-physical systems under adversarial attacks. IEEE Trans. Autom. control **59**(6), 1454–1467 (2014)

49. S. Fayyaz, M.M. Nazir, Handling security issues for smart grid applications using cloud computing framework. J. Emerg. Trends Comput. Inf. Sci. **3**(2), 285–287 (2012)

50. P.A. Forero, A. Cano, G.B. Giannakis, Consensus-based distributed support vector machines. J. Mach. Learn. Res. **11**(5), 1663–1707 (2010)

51. B.A. Francis, J.C. Doyle, Linear control theory with an H_∞ optimality criterion. SIAM J. Control Optim. **25**(4), 815–844 (1987)

52. A. Granas, J. Dugundji, *Fixed Point Theory* (Springer Science & Business Media, Berlin/Heidelberg, 2013)

53. V.C. Gungor, D. Sahin, T. Kocak, S. Ergut, C. Buccella, C. Cecati, G.P. Hancke, Smart grid technologies: communication technologies and standards. IEEE Trans. Ind. Inf. **7**(4), 529–539 (2011)

54. C.S.V. Gutiérrez, L.U.S. Juan, I.Z. Ugarte, V.M. Vilches, Time-sensitive networking for robotics (2018). Preprint. arXiv:1804.07643

55. F. Harrell, Regression Modeling Strategies. As Implemented in R Package rms Version, vol. 3(3) (2013)

56. J.C. Harsanyi, Games with incomplete information played by bayesian players, I–III part I. The basic model. Manage. Sci. **14**(3), 159–182 (1967)

57. J.C. Harsanyi, Games with incomplete information played by Bayesian players part II. Bayesian equilibrium points. Manage. Sci. **14**(5), 320–334 (1968)

58. W. Heemels, K.H. Johansson, P. Tabuada, An introduction to event-triggered and self-triggered control, in *2012 IEEE 51st IEEE Conference on Decision and Control (CDC)* (IEEE, New York, 2012), pp. 3270–3285

59. I. Hong, J. Byun, S. Park, Cloud computing-based building energy management system with zigbee sensor network, in *2012 Sixth International Conference on Innovative Mobile and Internet Services in Ubiquitous Computing* (IEEE, New York, 2012), pp. 547–551

60. K. Horák, B. Bošanskỳ, M. Pěchouček, Heuristic search value iteration for one-sided partially observable stochastic games, in *Thirty-First AAAI Conference on Artificial Intelligence* (2017)

61. K. Horák, Q. Zhu, B. Bošanskỳ, Manipulating adversary's belief: a dynamic game approach to deception by design for proactive network security, in *International Conference on Decision and Game Theory for Security* (Springer, New York, 2017), pp. 273–294

62. G. Hu, W.P. Tay, Y. Wen, Cloud robotics: architecture, challenges and applications. IEEE Network **26**(3), 21–28 (2012)

63. L. Huang, Q. Zhu, Analysis and computation of adaptive defense strategies against advanced persistent threats for cyber-physical systems, in *International Conference on Decision and Game Theory for Security* (Springer, New York, 2018), pp. 205–226

64. L. Huang, Q. Zhu, Adaptive strategic cyber defense for advanced persistent threats in critical infrastructure networks. ACM SIGMETRICS Perform. Eval. Rev. **46**(2), 52–56 (2019)

65. L. Huang, Q. Zhu, Dynamic bayesian games for adversarial and defensive cyber deception, in *Autonomous Cyber Deception* (Springer, New York, 2019), pp. 75–97

66. L. Huang, Q. Zhu, Dynamic games of asymmetric information for deceptive autonomous vehicles (2019)

67. L. Huang, Q. Zhu, A dynamic games approach to proactive defense strategies against advanced persistent threats in cyber-physical systems. Comput. Secur. **89**, 101660 (2020)

68. T. Hunter, T. Moldovan, M. Zaharia, S. Merzgui, J. Ma, M.J. Franklin, P. Abbeel, A.M. Bayen, Scaling the mobile millennium system in the cloud, in *Proceedings of the 2nd ACM Symposium on Cloud Computing* (2011), pp. 28

69. D. Hunziker, M. Gajamohan, M. Waibel, R. D'Andrea, Rapyuta: The roboearth cloud engine, in *2013 IEEE International Conference on Robotics and Automation* (IEEE, New York, 2013), pp. 438–444

70. O.C. Imer, S. Yüksel, T. Başar, Optimal control of lti systems over unreliable communication links. Automatica **42**(9), 1429–1439 (2006)

71. S. Jajodia, A.K. Ghosh, V. Swarup, C. Wang, X.S. Wang, *Moving Target Defense: Creating Asymmetric Uncertainty for Cyber Threats*, vol. 54 (Springer Science & Business Media, Berlin/Heidelberg, 2011)

72. S. Jajodia, V. Subrahmanian, V. Swarup, C. Wang, *Cyber Deception*, vol. 6 (Springer, New York, 2016)

73. J. Katz, Y. Lindell, *Introduction to Modern Cryptography* (CRC Press, Boca Raton, FL, 2014)

74. B. Kehoe, D. Berenson, K. Goldberg, Toward cloud-based grasping with uncertainty in shape: estimating lower bounds on achieving force closure with zero-slip push grasps, in *IEEE International Conference on Robotics and Automation (ICRA)* (2012), pp. 576–583

75. B. Kehoe, S. Patil, P. Abbeel, K. Goldberg, A survey of research on cloud robotics and automation. IEEE Trans. Autom. Sci. Eng. **12**(2), 398–409 (2015)

76. A.J. Kerns, D.P. Shepard, J.A. Bhatti, T.E. Humphreys, Unmanned aircraft capture and control via GPS spoofing. J. Field Robot. **31**(4), 617–636 (2014)

77. J.-S. Kim, T.-W. Yoon, A. Jadbabaie, C. De Persis, Input-to-state stabilizing mpc for neutrally stable linear systems subject to input constraints, in *43rd IEEE Conference on Decision and Control (CDC)*, vol. 5 (2004), pp. 5041–5046

78. H. Kim, Y.-J. Kim, K. Yang, M. Thottan, Cloud-based demand response for smart grid: architecture and distributed algorithms, in *2011 IEEE International Conference on Smart Grid Communications (SmartGridComm)* (IEEE, New York, 2011), pp. 398–403

79. L.A. Kirschgens, I.Z. Ugarte, E.G. Uriarte, A.M. Rosas, V.M. Vilches, Robot hazards: from safety to security (2018). Preprint. arXiv:1806.06681

80. K. Kogiso, T. Fujita, Cyber-security enhancement of networked control systems using homomorphic encryption, in *2015 IEEE 54th Annual Conference on Decision and Control (CDC)* (IEEE, New York, 2015), pp. 6836–6843

81. J. Kramer, M. Scheutz, Development environments for autonomous mobile robots: a survey. Autonom. Robots **22**(2), 101–132 (2007)

82. V. Krishnamurthy, *Partially Observed Markov Decision Processes* (Cambridge University Press, Cambridge, 2016)

83. P.R. Kumar, P. Varaiya, *Stochastic Systems: Estimation, Identification, and Adaptive Control* (SIAM, Philadelphia, PA, 2015)

84. H. Kwakernaak, R. Sivan, *Linear Optimal Control Systems*, vol. 1 (Wiley-Interscience, New York, 1972)

85. R.L. Lagendijk, Z. Erkin, M. Barni, Encrypted signal processing for privacy protection: conveying the utility of homomorphic encryption and multiparty computation. IEEE Signal Process. Mag. **30**(1), 82–105 (2013)

86. D. Lehmann, E. Henriksson, K.H. Johansson, Event-triggered model predictive control of discrete-time linear systems subject to disturbances, in *European Control Conference (ECC)* (2013), pp. 1156–1161

87. X. Lei, X. Liao, T. Huang, H. Li, C. Hu, Outsourcing large matrix inversion computation to a public cloud. IEEE Trans. Cloud Comput. **1**, 78–87 (2013)

88. X. Lei, X. Liao, T. Huang, H. Li, Cloud computing service: the case of large matrix determinant computation. IEEE Trans. Serv. Comput. **PP**, 688–700 (2014)

89. F. Li, A. Lai, D. Ddl, Evidence of advanced persistent threat: a case study of malware for political espionage, in *2011 6th International Conference on Malicious and Unwanted Software (MALWARE)* (IEEE, New York, 2011), pp. 102–109

90. H. Li, L. Lai, R.C. Qiu, A denial-of-service jamming game for remote state monitoring in smart grid, in *2011 45th Annual Conference on Information Sciences and Systems (CISS)*, pp. 1–6 (IEEE, New York, 2011)

91. Z. Li, C. Chen, K. Wang, Cloud computing for agent-based urban transportation systems. IEEE Intell. Syst. **26**(1), 73–79 (2011)

92. M. Li, D.G. Andersen, A.J. Smola, K. Yu, Communication efficient distributed machine learning with the parameter server, in *Advances in Neural Information Processing Systems* (2014), pp. 19–27

93. Y. Li, D.E. Quevedo, V. Lau, L. Shi, Multi-sensor transmission power scheduling for remote state estimation under sinr model, in *2014 IEEE 53rd Annual Conference on Decision and Control (CDC)* (IEEE, New York, 2014), pp. 1055–1060

94. Y. Li, L. Shi, P. Cheng, J. Chen, D.E. Quevedo, Jamming attacks on remote state estimation in cyber-physical systems: a game-theoretic approach. IEEE Trans. Autom. Control **60**(10), 2831–2836 (2015)

95. D. Liberzon, *Switching in Systems and Control* (Springer Science & Business Media, Berlin/Heidelberg, 2012)

96. Y. Liu, P. Ning, M.K. Reiter, False data injection attacks against state estimation in electric power grids. ACM Trans. Inf. Syst. Secur. **14**(1), 13 (2011)

97. I. Lopez, M. Aguado, Cyber security analysis of the european train control system. Commun. Mag. IEEE **53**(10), 110–116 (2015)

98. D.G. Luenberger, *Optimization by Vector Space Methods* (Wiley, New York, 1997)

99. W. Luo, T. Hu, C. Zhang, Y. Wei, Digital twin for CNC machine tool: modeling and using strategy. J. Amb. Intell. Human. Comput. **10**(3), 1129–1140 (2019)

100. A. Mahimkar, V. Shmatikov, Game-based analysis of denial-of-service prevention protocols, in *18th IEEE Computer Security Foundations Workshop (CSFW'05)* (IEEE, New York, 2005), pp. 287–301

101. M.H. Manshaei, Q. Zhu, T. Alpcan, T. Bacşar, J.-P. Hubaux, Game theory meets network security and privacy. ACM Comput. Surv. **45**(3), 25 (2013)

102. D.Q. Mayne, J.B. Rawlings, C.V. Rao, P.O. Scokaert, Constrained model predictive control: stability and optimality. Automatica **36**(6), 789–814 (2000)

103. J. McClean, C. Stull, C. Farrar, D. Mascareñas, A preliminary cyber-physical security assessment of the robot operating system (ROS), in *Unmanned Systems Technology XV*, vol. 8741 (International Society for Optics and Photonics, Bellingham, WA, 2013), pp. 874110

104. R.K. Mehra, J. Peschon, An innovations approach to fault detection and diagnosis in dynamic systems. Automatica **7**(5), 637–640 (1971)
105. C. Meng, T. Wang, W. Chou, S. Luan, Y. Zhang, Z. Tian, Remote surgery case: robot-assisted teleneurosurgery, in *IEEE International Conference on Robotics and Automation (ICAR)* (2004), pp. 819–823
106. R.C. Merkle, A certified digital signature, in *Conference on the Theory and Application of Cryptology* (Springer, New York, 1989), pp. 218–238
107. F. Miao, Q. Zhu, A moving-horizon hybrid stochastic game for secure control of cyber-physical systems, in *2014 IEEE 53rd Annual Conference on Decision and Control (CDC)* (IEEE, New York, 2014), pp. 517–522
108. F. Miao, Q. Zhu, M. Pajic, G.J. Pappas, Coding schemes for securing cyber-physical systems against stealthy data injection attacks. IEEE Trans. Control Netw. Syst. **4**(1), 106–117 (2017)
109. B. Miller, D. Rowe, A survey scada of and critical infrastructure incidents, in *Proceedings of the 1st Annual Conference on Research in Information Technology* (ACM, New York, 2012), pp. 51–56
110. Y. Mo, B. Sinopoli, Integrity attacks on cyber-physical systems, in *Proceedings of the 1st International Conference on High Confidence Networked Systems* (ACM, New York, 2012), pp. 47–54
111. Y. Mo, B. Sinopoli, On the performance degradation of cyber-physical systems under stealthy integrity attacks. IEEE Trans. Autom. Control **61**(9), 2618–2624 (2015)
112. A. Mohammadi, M.H. Manshaei, M.M. Moghaddam, Q. Zhu, A game-theoretic analysis of deception over social networks using fake avatars, in *International Conference on Decision and Game Theory for Security* (Springer, New York, 2016), pp. 382–394
113. J. Moreno, J.M. Riera, L. De Haro, C. Rodriguez, A survey on future railway radio communications services: challenges and opportunities. Commun. Mag. IEEE **53**(10), 62–68 (2015)
114. O. Morgenstern, J. Von Neumann, *Theory of Games and Economic Behavior* (Princeton University Press, Princeton, 1953)
115. E.R. Naru, H. Saini, M. Sharma, A recent review on lightweight cryptography in IoT, in *2017 International Conference on I-SMAC (IoT in Social, Mobile, Analytics and Cloud)(I-SMAC)* (IEEE, New York, 2017), pp. 887–890
116. M. Nasri, M. Kargahi, A method for improving delay-sensitive accuracy in real-time embedded systems, in *2012 IEEE 18th International Conference on Embedded and Real-Time Computing Systems and Applications (RTCSA)* (IEEE, New York, 2012), pp. 378–387
117. C. Neuman, Challenges in security for cyber-physical systems, in *DHS: S&T Workshop on Future Directions in Cyber-Physical Systems Security*, vol. 7. Citeseer (2009)
118. J.M. O'Kane, A gentle introduction to ROS (2014)
119. G. Owen, *Game Theory*, 3rd edn. (Academic Press, New York, 1995)
120. M. Pajic, I. Lee, G.J. Pappas, Attack-resilient state estimation for noisy dynamical systems. IEEE Trans. Control Netw. Syst. **4**(1), 82–92 (2016)
121. P. Pandey, D. Pompili, J. Yi, Dynamic collaboration between networked robots and clouds in resource-constrained environments. IEEE Trans. Autom. Sci. Eng. **12**(2), 471–480 (2015)
122. A. Papoulis, S.U. Pillai, *Probability, Random Variables, and Stochastic Processes* (Tata McGraw-Hill Education, New York, 2002)
123. F. Pasqualetti, F. Dorfler, F. Bullo, Control-theoretic methods for cyberphysical security: geometric principles for optimal cross-layer resilient control systems. IEEE Control Syst. Mag. **35**(1), 110–127 (2015)
124. J. Pawlick, Q. Zhu, Proactive defense against physical denial of service attacks using poisson signaling games, in *International Conference on Decision and Game Theory for Security* (Springer, New York, 2017), pp. 336–356
125. J. Pawlick, Q. Zhu, Strategic trust in cloud-enabled cyber-physical systems with an application to glucose control. IEEE Trans. Inf. Foren. Secur. **12**(12), 2906–2919 (2017)
126. J. Pawlick, S. Farhang, Q. Zhu, Flip the cloud: cyber-physical signaling games in the presence of advanced persistent threats, in *International Conference on Decision and Game Theory for Security* (Springer, New York, 2015), pp. 289–308

127. J. Pawlick, J. Chen, Q. Zhu, iSTRICT: an interdependent strategic trust mechanism for the cloud-enabled internet of controlled things. IEEE Trans. Inf. Foren. Secur. **14**(6), 1654–1669 (2018)

128. J. Pawlick, E. Colbert, Q. Zhu, Modeling and analysis of leaky deception using signaling games with evidence. IEEE Trans. Inf. Foren. Secur. **14**(7), 1871–1886 (2018)

129. J. Pawlick, E. Colbert, Q. Zhu, A game-theoretic taxonomy and survey of defensive deception for cybersecurity and privacy. ACM Comput. Surv. **52**(4), 1–28 (2019)

130. J.M. Porta, N. Vlassis, M.T. Spaan, P. Poupart, Point-based value iteration for continuous POMDPs. J. Mach. Learn. Res. **7**, 2329–2367 (2006)

131. P. Poupart, Approximate value-directed belief state monitoring for partially observable Markov decision processes. PhD thesis, University of British Columbia, 2000

132. M. Quigley, K. Conley, B. Gerkey, J. Faust, T. Foote, J. Leibs, R. Wheeler, A.Y. Ng, ROS: an open-source robot operating system, in *ICRA Workshop on Open Source Software*, vol. 3 (Kobe, Japan, 2009), p. 5

133. J. Rasmussen, Patching frequency best practices (2018)

134. S. Rass, *Cyber-Security in Critical Infrastructures: A Game-Theoretic Approach* (Springer Nature, Cham, 2020)

135. S. Rass, A. Alshawish, M.A. Abid, S. Schauer, Q. Zhu, H. De Meer, Physical intrusion games–optimizing surveillance by simulation and game theory. IEEE Access **5**, 8394–8407 (2017)

136. M. Ratto, R. Ree, Materializing information: 3d printing and social change. First Monday **17**(7) (2012). https://doi.org/10.5210/fm.v17i7.3968

137. C. Reiger, I. Ray, Q. Zhu, M.A. Haney, Industrial control systems security and resiliency. *Practice and Theory* (Springer, Cham, 2019)

138. L. Riazuelo, M. Tenorth, D. Di Marco, M. Salas, D. Gálvez-López, L. Mösenlechner, L. Kunze, M. Beetz, J.D. Tardós, L. Montano, et al., Roboearth semantic mapping: a cloud enabled knowledge-based approach. IEEE Trans. Autom. Sci. Eng. **12**(2), 432–443 (2015)

139. M. Ribeiro, K. Grolinger, M.A. Capretz, MLaaS: machine learning as a service, in *2015 IEEE 14th International Conference on Machine Learning and Applications (ICMLA)* (IEEE, New York, 2015), pp. 896–902

140. C. Rieger, Q. Zhu, A hierarchical multi-agent dynamical system architecture for resilient control systems, in *2013 6th International Symposium on Resilient Control Systems (ISRCS)* (IEEE, New York, 2013), pp. 6–12

141. C.G. Rieger, D.I. Gertman, M.A. McQueen, Resilient control systems: next generation design research, in *2009 2nd Conference on Human System Interactions* (IEEE, New York, 2009), pp. 632–636

142. C. Rieger, Q. Zhu, T. Başar, Agent-based cyber control strategy design for resilient control systems: concepts, architecture and methodologies, in *2012 5th International Symposium on Resilient Control Systems (ISRCS)* (IEEE, New York, 2012), pp. 40–47

143. Y.E. Sagduyu, A. Ephremides, A game-theoretic analysis of denial of service attacks in wireless random access. Wirel. Netw. **15**(5), 651–666 (2009)

144. O. Saha, P. Dasgupta, A comprehensive survey of recent trends in cloud robotics architectures and applications. Robotics **7**(3), 47 (2018)

145. R. Saltzman, A. Sharabani, Active man in the middle attacks. OWASP AU (2009)

146. H. Sandberg, S. Amin, K.H. Johansson, Cyberphysical security in networked control systems: an introduction to the issue. IEEE Control Syst. **35**(1), 20–23 (2015)

147. C. Schifers, G. Hans, IEEE Standard for Communications-Based Train Control (cbtc) performance and functional requirements, in *Vehicular Technology Conference Proceedings, VTC*, (2000), pp. 1581–1585

148. P. Seiler, R. Sengupta, Analysis of communication losses in vehicle control problems, in *Proceedings of the 2001 American Control Conference*, vol. 2 (2001), pp. 1491–1496

149. G. Shani, J. Pineau, R. Kaplow, A survey of point-based POMDP solvers. Auton. Agents Multi-Agent Syst. **27**(1), 1–51 (2013)

150. S.K. Sharma, X. Wang, Toward massive machine type communications in ultra-dense cellular iot networks: current issues and machine learning-assisted solutions. IEEE Commun. Surv. Tutor. **22**(1), 426–471 (2019)

151. S. Shen, Y. Li, H. Xu, Q. Cao, Signaling game based strategy of intrusion detection in wireless sensor networks. Comput. Math. Appl. **62**(6), 2404–2416 (2011)

152. V. Shnayder, M. Hempstead, B.-R. Chen, G.W. Allen, M. Welsh, Simulating the power consumption of large-scale sensor network applications, in *Proceedings of the 2nd International Conference on Embedded Networked Sensor Systems* (ACM, New York, 2004), pp. 188–200

153. Y. Simmhan, S. Aman, B. Cao, M. Giakkoupis, A. Kumbhare, Q. Zhou, D. Paul, C. Fern, A. Sharma, V.K. Prasanna, An informatics approach to demand response optimization in smart grids. Technical report, City of Los Angeles Department, 2011

154. Y. Simmhan, A.G. Kumbhare, B. Cao, V. Prasanna, An analysis of security and privacy issues in smart grid software architectures on clouds, in *2011 IEEE 4th International Conference on Cloud Computing* (IEEE, New York, 2011), pp. 582–589

155. E.J. Sondik, The optimal control of partially observable Markov processes. Technical report, Stanford Electronics Labs (SEL), Stanford University, Technical Report No. SU-SEL-71-017, 1971

156. E.J. Sondik, The optimal control of partially observable Markov processes over the infinite horizon: discounted costs. Oper. Res. **26**(2), 282–304 (1978)

157. J.B. Song, Q. Zhu, Performance of dynamic secure routing game, in *Game Theory for Networking Applications* (Springer, New York, 2019), pp. 37–56

158. A. Sternstein, Things can go kaboom when a defense contractor's 3D printer get hacked (2014)

159. L. Sturm, C. Williams, J. Camelio, J. White, R. Parker, Cyber-physical vunerabilities in additive manufacturing systems. Context **7**, 8 (2014)

160. C. Tankard, Advanced persistent threats and how to monitor and deter them. Netw. Secur. **2011**(8), 16–19 (2011)

161. F. Tao, Y. Hu, L. Zhang, Theory and Practice: Optimal Resource Service Allocation in Manufacturing Grid (China Machine Press, Beijing, 2010)

162. F. Tao, L. Zhang, V. Venkatesh, Y. Luo, Y. Cheng, Cloud manufacturing: a computing and service-oriented manufacturing model. Proc. Inst. Mech. Eng. Part B J. Eng. Manuf. **225**(10), 1969–1976 (2011)

163. F. Tao, J. Cheng, Q. Qi, M. Zhang, H. Zhang, F. Sui, Digital twin-driven product design, manufacturing and service with big data. Int. J. Adv. Manuf. Technol. **94**(9–12), 3563–3576 (2018)

164. S. Tatikonda, S. Mitter, Control under communication constraints. IEEE Trans. Autom. Control **49**(7), 1056–1068 (2004)

165. A. Teixeira, D. Pérez, H. Sandberg, K.H. Johansson, Attack models and scenarios for networked control systems, in *Proceedings of the 1st International Conference on High Confidence Networked Systems* (ACM, New York, 2012), pp. 55–64

166. L. Turnbull, B. Samanta, Cloud robotics: formation control of a multi robot system utilizing cloud infrastructure, in *Southeastcon, 2013 Proceedings of IEEE* (2013)

167. M. Van Dijk, A. Juels, A. Oprea, R.L. Rivest, Flipit: The game of stealthy takeover. J. Cryptol. **26**(4), 655–713 (2013)

168. A.N. Venkat, I. Hiskens, J.B. Rawlings, S.J. Wright, et al., Distributed MPC strategies with application to power system automatic generation control. IEEE Trans. Control Syst. Technol. **16**(6), 1192–1206 (2008)

169. H. Von Stackelberg, *Market Structure and Equilibrium* (Springer Science & Business Media, Berlin/Heidelberg, 2010)

170. C. Wang, K. Ren, J. Wang, Secure and practical outsourcing of linear programming in cloud computing, in *2011 Proceedings IEEE INFOCOM* (2011), pp. 820–828

171. H. Wang, F.R. Yu, L. Zhu, T. Tang, B. Ning, A cognitive control approach to communication-based train control systems. IEEE Trans. Intell. Transp. Syst. **16**(4), 1676–1689 (2015)

172. B.M. Weiss,   Closed-loop control of a 3D printer gantry.   PhD thesis, University of Washington, 2014
173. X. Xu, From cloud computing to cloud manufacturing. Robot. Comput.-Integr. Manuf. **28**(1), 75–86 (2012)
174. Z. Xu, A. Easwaran, A game-theoretic approach to secure estimation and control for cyber-physical systems with a digital twin, in *2020 ACM/IEEE 11th International Conference on Cyber-Physical Systems (ICCPS)* (IEEE, New York, 2020), pp. 20–29
175. Z. Xu, Q. Zhu,   A cyber-physical game framework for secure and resilient multi-agent autonomous systems, in *2015 IEEE 54th Annual Conference on Decision and Control (CDC)* (IEEE, New York, 2015), pp. 5156–5161
176. Z. Xu, Q. Zhu,   Environment-aware power generation scheduling in smart grids, in *2015 IEEE International Conference on Smart Grid Communications (SmartGridComm)* (IEEE, New York, 2015), pp. 253–258
177. Z. Xu, Q. Zhu,   Secure and resilient control design for cloud enabled networked control systems, in *Proceedings of the First ACM Workshop on Cyber-Physical Systems-Security and/or PrivaCy* (ACM, New York, 2015), pp. 31–42
178. Z. Xu, Q. Zhu,   Cross-layer secure and resilient control of delay-sensitive networked robot operating systems, in *2018 IEEE Conference on Control Technology and Applications (CCTA)* (IEEE, New York, 2018), pp. 1712–1717
179. M. Yampolskiy, T.R. Andel, J.T. McDonald, W.B. Glisson, A. Yasinsac,   Intellectual property protection in additive layer manufacturing: requirements for secure outsourcing, in *Proceedings of the 4th Program Protection and Reverse Engineering Workshop* (ACM, New York, 2014), p. 7
180. Q. Yang, Y. Liu, T. Chen, Y. Tong,   Federated machine learning: concept and applications. ACM Trans. Intell. Syst. Technol. **10**(2), 1–19 (2019)
181. X. Yao, Z. Chen, Y. Tian, A lightweight attribute-based encryption scheme for the internet of things. Fut. Gener. Comput. Syst. **49**, 104–112 (2015)
182. Y. Yuan, Q. Zhu, F. Sun, Q. Wang, T. Başar,   Resilient control of cyber-physical systems against denial-of-service attacks, in *2013 6th International Symposium on Resilient Control Systems (ISRCS)* (IEEE, New York, 2013), pp. 54–59
183. L. Zadeh, C. Desoer,   *Linear System Theory: The State Space Approach* (Courier Dover Publications, Mineola, New York, 2008)
184. A. Zanella, N. Bui, A. Castellani, L. Vangelista, M. Zorzi, Internet of things for smart cities. IEEE Internet Things J. **1**(1), 22–32 (2014)
185. M. Zanon, J.V. Frasch, M. Vukov, S. Sager, M. Diehl, Model predictive control of autonomous vehicles, in *Optimization and Optimal Control in Automotive Systems* (Springer, New York, 2014), pp. 41–57
186. W. Zeng, M.-Y. Chow,   Optimal tradeoff between performance and security in networked control systems based on coevolutionary algorithms. IEEE Trans. Ind. Electron. **59**(7), 3016–3025 (2012)
187. R. Zhang, Q. Zhu,   Secure and resilient distributed machine learning under adversarial environments, in *2015 18th International Conference on Information Fusion (Fusion)* (IEEE, New York, 2015), pp. 644–651
188. R. Zhang, Q. Zhu, A game-theoretic analysis of label flipping attacks on distributed support vector machines, in *2017 51st Annual Conference on Information Sciences and Systems (CISS)* (IEEE, New York, 2017), pp. 1–6
189. R. Zhang, Q. Zhu,   A game-theoretic defense against data poisoning attacks in distributed support vector machines, in *2017 IEEE 56th Annual Conference on Decision and Control (CDC)* (IEEE, New York, 2017), pp. 4582–4587
190. T. Zhang, Q. Zhu, Dynamic differential privacy for ADMM-based distributed classification learning. IEEE Trans. Inf. Foren. Secur. **12**(1), 172–187 (2017)
191. T. Zhang, Q. Zhu,   Strategic defense against deceptive civilian GPS spoofing of unmanned aerial vehicles, in *International Conference on Decision and Game Theory for Security* (Springer, New York, 2017), pp. 213–233

192. T. Zhang, Q. Zhu, Distributed privacy-preserving collaborative intrusion detection systems for vanets. IEEE Trans. Signal Inf. Process. Over Netw. **4**(1), 148–161 (2018)
193. R. Zhang, Q. Zhu, A game-theoretic approach to design secure and resilient distributed support vector machines. IEEE Trans. Neural Netw. Learn. Syst. **29**(11), 5512–5527 (2018)
194. R. Zhang, Q. Zhu, Consensus-based transfer linear support vector machines for decentralized multi-task multi-agent learning, in *2018 52nd Annual Conference on Information Sciences and Systems (CISS)* (IEEE, New York, 2018), pp. 1–6
195. Q. Zhang, L. Cheng, R. Boutaba, Cloud computing: state-of-the-art and research challenges. J. Internet Serv. Appl. **1**(1), 7–18 (2010)
196. Q. Zhang, Q. Zhu, M.F. Zhani, R. Boutaba, J.L. Hellerstein, Dynamic service placement in geographically distributed clouds. IEEE J. Select. Areas Commun. **31**(12), 762–772 (2013)
197. T. Zhang, L. Huang, J. Pawlick, Q. Zhu, Game-theoretic analysis of cyber deception: evidence-based strategies and dynamic risk mitigation, in *Modeling and Design of Secure Internet of Things* (Wiley, New York, 2020), pp. 27–58
198. G. Zhao, C. Rong, J. Li, F. Zhang, Y. Tang, Trusted data sharing over untrusted cloud storage providers, in *2010 IEEE Second International Conference on Cloud Computing Technology and Science (CloudCom)* (IEEE, New York, 2010), pp. 97–103
199. Q. Zhu, Control challenges for resilient control systems (2020). Preprint. arXiv:2001.00712
200. Q. Zhu, T. Başar, Dynamic policy-based IDS configuration, in *Proceedings of the 48h IEEE Conference on Decision and Control (CDC) Held Jointly with 2009 28th Chinese Control Conference* (IEEE, New York, 2009), pp. 8600–8605
201. Q. Zhu, T. Başar, Robust and resilient control design for cyber-physical systems with an application to power systems, in *2011 50th IEEE Conference on Decision and Control and European Control Conference (CDC-ECC)* (IEEE, New York, 2011), pp. 4066–4071
202. Q. Zhu, T. Başar, A dynamic game-theoretic approach to resilient control system design for cascading failures, in *Proceedings of the 1st International Conference on High Confidence Networked Systems* (ACM, New York, 2012), pp. 41–46
203. Q. Zhu, T. Başar, Game-theoretic approach to feedback-driven multi-stage moving target defense, in *International Conference on Decision and Game Theory for Security* (Springer, New York, 2013), pp. 246–263
204. Q. Zhu, T. Basar, Game-theoretic methods for robustness, security, and resilience of cyberphysical control systems: games-in-games principle for optimal cross-layer resilient control systems. Control Syst. IEEE **35**(1), 46–65 (2015)
205. Q. Zhu, L. Bushnell, Networked cyber-physical systems: interdependence, resilience and information exchange, in *2013 51st Annual Allerton Conference on Communication, Control, and Computing (Allerton)* (IEEE, New York, 2013), pp. 763–769
206. Q. Zhu, S. Rass, On multi-phase and multi-stage game-theoretic modeling of advanced persistent threats. IEEE Access **6**, 13958–13971 (2018)
207. Q. Zhu, H. Li, Z. Han, T. Başar, A stochastic game model for jamming in multi-channel cognitive radio systems, in *2010 IEEE International Conference on Communications* (IEEE, New York, 2010), pp. 1–6
208. Q. Zhu, H. Tembine, T. Başar, Network security configurations: a nonzero-sum stochastic game approach, in *Proceedings of the 2010 American Control Conference* (IEEE, New York, 2010), pp. 1059–1064
209. Q. Zhu, W. Saad, Z. Han, H.V. Poor, T. Başar, Eavesdropping and jamming in next-generation wireless networks: a game-theoretic approach, in *2011-MILCOM 2011 Military Communications Conference* (IEEE, New York, 2011), pp. 119–124
210. L. Zhu, F.R. Yu, B. Ning, T. Tang, Cross-layer handoff design in MIMO-enabled WLANs for communication-based train control (CBTC) systems. IEEE J. Sel. Areas Commun. **30**(4), 719–728 (2012)

211. Q. Zhu, A. Clark, R. Poovendran, T. Başar, Deceptive routing games, in *2012 IEEE 51st IEEE Conference on Decision and Control (CDC)* (IEEE, New York, 2012), pp. 2704–2711
212. Q. Zhu, Z. Yuan, J.B. Song, Z. Han, T. Basar, Interference aware routing game for cognitive radio multi-hop networks. IEEE J. Sel. Areas Commun. **30**(10), 2006–2015 (2012)
213. S. Zlobec, *Stable Parametric Programming*, vol. 57 (Springer Science & Business Media, Berlin/Heidelberg, 2013)

Index

© The Editor(s) (if applicable) and The Author(s), under exclusive license to
Springer Nature Switzerland AG 2020
Q. Zhu, Z. Xu, *Cross-Layer Design for Secure and Resilient Cyber-Physical Systems*,
Advances in Information Security 81, https://doi.org/10.1007/978-3-030-60251-2

Printed in the United States
by Baker & Taylor Publisher Services